Actively Caring for Safety

Actively Caring for Safety: The Psychological Science of Injury Prevention outlines proactive applications of applied behavioral science and humanism (i.e., humanistic behaviorism) for improving health and safety. This text provides evidence-based principles for customizing effective processes for improving the human dynamics of safety and health in various locations—from home to the workplace and throughout a community. World-renowned health/safety researcher, teacher, and consultant E. Scott Geller combines theory and principles in practical step-by-step procedures with behavioral science methods capable of enhancing safety awareness, reducing at-risk behavior, and facilitating ongoing participation in safety-related activities.

Drawing upon his bestselling works *Working Safe* and *The Psychology of Safety Handbook*, this book presents a science-based and practical approach to improving attitudes and behavior for achieving an injury-free work environment. The text has been improved and updated throughout and includes additional material on a rationale for language to replace common safety-related words that stifle human engagement. Plus, critical safety-relevant information is provided on empathy, emotional intelligence, self-motivation, positive psychology, psychological safety, the dramatic benefits of promoting perceptions of personal choice, and critical distinctions between leadership and management for optimizing workplace safety and productivity.

Written in an enjoyable, anecdotal, and engaging style, this is an essential read for any student, academic, researcher, or professional of health and safety.

"This book has been written by one of the world's leading experts. Dr. Geller's style is very accessible. People know his work better than any safety writer out there."

Dr. John Austin, *Reaching Results LLC, Kalamazoo, Michigan, USA*

"This book should have appeal anywhere there is heavy industry and a desire to keep employees safe. It is detailed, informative, and backed up by science in a way to engage experienced professionals or academics, but straightforward and practical enough to engage line-level employees and safety team members."

Dr. Steve Roberts, *Safety Performance Solutions, Inc, Blacksburg, VI, USA*

"Dr. Geller clearly articulates many decades of human behavior research in this work. Leaders and safety professionals can influence substantial safety and risk improvement by utilizing the concepts presented in this book."

Joseph E. Bolduc, *OnPoint Group, LLC, Daytona Beach, Florida, USA*

"The principles masterfully described in this classic book have stood the test of time as they have been used by hundreds of companies to bolster the safety culture of their workforce. This book provides clear tactics to diagnose and intervene on the issues presenting barriers to the safe performance of industrial organizations. First time readers will find this material as a 'disruptive' perspective on safety certainly ending up as an often-pulled addition to the safety professionals' bookshelf."

Dr. Tim Ludwig, *Appalachian State University, North Carolina, USA*

Actively Caring for Safety
The Psychological Science of Injury Prevention

Third Edition

E. Scott Geller
Alumni Distinguished Professor
Center for Applied Behavior Systems
Department of Psychology
Virginia Tech

CRC Press
Taylor & Francis Group
Boca Raton London New York

CRC Press is an imprint of the
Taylor & Francis Group, an **informa** business

Designed cover image: Nancy Poes

Third edition published 2024
by CRC Press
2385 NW Executive Center Drive, Suite 320, Boca Raton FL 33431

and by CRC Press
4 Park Square, Milton Park, Abingdon, Oxon, OX14 4RN

CRC Press is an imprint of Taylor & Francis Group, LLC

© 2024 E. Scott Geller

First edition published by CRC Press 1996
Second edition published by CRC Press 2001

Reasonable efforts have been made to publish reliable data and information, but the author and publisher cannot assume responsibility for the validity of all materials or the consequences of their use. The authors and publishers have attempted to trace the copyright holders of all material reproduced in this publication and apologize to copyright holders if permission to publish in this form has not been obtained. If any copyright material has not been acknowledged please write and let us know so we may rectify in any future reprint.

Except as permitted under U.S. Copyright Law, no part of this book may be reprinted, reproduced, transmitted, or utilized in any form by any electronic, mechanical, or other means, now known or hereafter invented, including photocopying, microfilming, and recording, or in any information storage or retrieval system, without written permission from the publishers.

For permission to photocopy or use material electronically from this work, access www.copyright.com or contact the Copyright Clearance Center, Inc. (CCC), 222 Rosewood Drive, Danvers, MA 01923, 978–750–8400. For works that are not available on CCC please contact mpkbookspermissions@tandf.co.uk

Trademark notice: Product or corporate names may be trademarks or registered trademarks and are used only for identification and explanation without intent to infringe.

Library of Congress Cataloging-in-Publication Data
Names: Geller, E. Scott, 1942– author.
Title: Actively caring for safety : the psychological science of injury prevention /
  E. Scott Geller, alumni distinguished professor, Center for Applied Behavior
  Systems, Department of Psychology, Virginia Tech.
Description: Third edition. | Boca Raton, FL : CRC Press, 2024. | Includes
  bibliographical references and index.
Identifiers: LCCN 2023056083 (print) | LCCN 2023056084 (ebook) |
  ISBN 9781032572611 (hbk) | ISBN 9781032566900 (pbk) |
  ISBN 9781003438564 (ebk)
Subjects: LCSH: Industrial safety—Psychological aspects. | Corporate culture.
Classification: LCC T55.3.B43 G447 2024 (print) | LCC T55.3.B43 (ebook) |
  DDC 658.3/82—dc23/eng/20240213
LC record available at https://lccn.loc.gov/2023056083
LC ebook record available at https://lccn.loc.gov/2023056084

ISBN: 978-1-032-57261-1 (hbk)
ISBN: 978-1-032-56690-0 (pbk)
ISBN: 978-1-003-43856-4 (ebk)

DOI: 10.1201/9781003438564

Typeset in Times LT Std
by Apex CoVantage, LLC

Dedication

Past

to my mother (Margaret J. Scott, R.N.) and father (Edward I. Geller, M.D.), who taught me the value of learning and reinforced my never-ending need to achieve.

to B. F. Skinner and W. Edwards Deming, who developed and researched the most applicable principles in this text and inspired me to teach them.

Present

to my daughters Krista S. Geller and Karly S. Geller, who inspire me to keep researching, teaching, and writing as my legacy for their make-a-difference lives.

to the students and associates in our Virginia Tech (VT) Center for Applied Behavior Systems, whose data collection and analyses provided empirical evidence for many of the practical interventions exemplified in this text.

Future

to my grandchildren (Harrison, Hattie, and Huxley Coulehan), who will learn the AC4P principles from their mom—Krista S. Geller—and will achieve successive merits of achievement, culminating with experiencing the sense of accomplishment I feel by completing this scholarship.

to the Advisory Board and the Board of Directors for the AC4P Foundation, Inc—Evan Alvarez, Samuel Browning, Carol Geller, Karly Geller, Krista Geller, John Hodge, Loralee Hoffer, Bobby Kipper, Jerald Monahan, Denise Murray, Tyler Parker-Rollins, Terri Reynolds, Dolores Robinson, Anastasia Semenova, Shannon Snapp, Rev. J. R. Thickland, Michael Thompson, Jack Wardale, and their colleagues—who will disseminate the AC4P principles and practices revealed in this text on a large scale in order to improve safety, health, and human welfare in organizations, educational institutions, and communities worldwide.

Contents

Preface xvii
Acknowledgments xxiii
About the Author xxvii

PART ONE Orientation and Alignment

Chapter 1 Choosing the Right Approach 3

    1.1 Selecting the Best Approach 3
        1.1.1 Behavior-Based Programs 4
        1.1.2 Comprehensive Ergonomics 4
        1.1.3 Engineering Changes 5
        1.1.4 Group Problem Solving 5
        1.1.5 Government Action (in Finland) 5
        1.1.6 Management Audits 5
        1.1.7 Stress Management 5
        1.1.8 Poster Campaigns 6
        1.1.9 Personnel Selection 6
        1.1.10 Near Miss* Reporting 6
    1.2 The Critical Human Element 6
    1.3 The Folly of Choosing What Sounds Good 8
    1.4 Relying on Research 9
    1.5 Start with Behavior 10
    1.6 In Summary 11

Chapter 2 Starting with Theory 12

    2.1 The Mission Statement 12
    2.2 Theory as a Map 12
    2.3 Relevance to OSH 14
    2.4 A Basic Mission and Theory 15
    2.5 Behavior-Based vs. Person-Based Approaches 16
    2.6 The Person-Based Approach 16
    2.7 The Behavior-Based Approach 17
    2.8 Considering Cost Effectiveness 18
        2.8.1 Integrating Approaches 19
    2.9 In Summary 20

Chapter 3 Paradigm Shifts for a Total Safety Culture 22

    3.1 The Old Three Es 22
        3.1.1 Three Other Es 23
    3.2 Shifting Paradigms 24
        3.2.1 From Government Regulation to Corporate Responsibility 24
        3.2.2 From Failure Focused to Achievement Focused 25
        3.2.3 From Outcome Focused to Behavior Focused 26

    3.2.4 From Top-Down Control to Bottom-Up Involvement 27
    3.2.5 From Rugged Individualism to Interdependent Teamwork 27
    3.2.6 From a Piecemeal to a Systems Approach 28
    3.2.7 From Fault Finding to Fact Finding 28
    3.2.8 From Reactive to Proactive 30
    3.2.9 From Quick Fix to Continuous Improvement 30
    3.2.10 From Priority to Value 30
    3.2.11 Enduring Values 31
  3.3 In Summary 31

PART TWO Human Barriers to Safety

Chapter 4 The Complexity of People 35
  4.1 Fighting Human Nature 35
    4.1.1 Dimensions of Human Nature 36
  4.2 Cognitive Failures 38
    4.2.1 Capture Errors 38
    4.2.2 Description Errors 39
    4.2.3 Loss-of-Activation Errors 39
    4.2.4 Mode Errors 39
    4.2.5 Mistakes and Calculated Risks 40
  4.3 Interpersonal Factors 40
    4.3.1 Peer Influence 40
    4.3.2 Power of Authority 41
  4.4 In Summary 43

Chapter 5 Sensation, Perception, and Perceived Risk 44
  5.1 Selective Sensation or Perception 44
    5.1.1 Biased by Context 46
    5.1.2 Biased by Our Past 47
    5.1.3 Relevance to Achieving a TSC 48
  5.2 Perceived Risk 48
    5.2.1 Real vs. Perceived Risk 49
    5.2.2 The Power of Perceived Choice 50
    5.2.3 Familiarity Breeds Complacency 50
    5.2.4 Sympathy for Victims 50
    5.2.5 Understood and Controllable Hazards 51
    5.2.6 Acceptable Consequences 51
    5.2.7 Sense of Fairness 52
    5.2.8 Risk Compensation 52
    5.2.9 Implications of Risk Compensation 53
    5.2.10 Critical Impact of Perceived Choice 53
  5.3 In Summary 54

Chapter 6 Stress vs. Distress 56
  6.1 What Is Stress? 57
    6.1.1 Constructive or Destructive? 58

Contents

	6.1.2	The Eyes of the Beholder	59
6.2		Identifying Stressors	60
6.3		Coping with Stressors	61
	6.3.1	Person Factors	62
	6.3.2	Fit for Stressors	63
	6.3.3	Perceived Choice	64
	6.3.4	Social Factors	66
6.4		Attributional Bias	67
	6.4.1	The Fundamental Attribution Error	67
	6.4.2	The Self-Serving Bias	68
6.5		In Summary	68

PART THREE Behavior-Based Psychology

Chapter 7 Basic Principles 71

7.1	Primacy of Behavior	71
	7.1.1 Reducing At-Risk Behaviors	72
	7.1.2 Increasing Safe Behaviors	74
7.2	Learning from Experience	75
	7.2.1 Classical Conditioning	75
	7.2.2 Operant Conditioning	76
	7.2.3 Observational Learning	78
	7.2.4 Overlapping Types of Learning	79
7.3	In Summary	80

Chapter 8 Identifying Critical Behaviors 82

8.1	The DO IT Process	82
8.2	Defining Target Behaviors	84
	8.2.1 What Is Behavior?	85
	8.2.2 Describing Behaviors	85
	8.2.3 Multiple Behaviors	85
8.3	Observing Behavior	86
	8.3.1 A Personal Example	87
	8.3.2 Using the Critical Behavior Checklist	89
8.4	Two Basic Approaches	89
	8.4.1 Starting Small	90
	8.4.2 Observing Multiple Behaviors	91
8.5	In Summary	92

Chapter 9 Behavioral Safety Analysis 93

9.1	Reducing Behavioral Discrepancy	93
	9.1.1 Can the Task Be Simplified?	93
	9.1.2 Is a Quick Fix Available?	95
	9.1.3 Is Safe Behavior Punished?	95
	9.1.4 Is At-Risk Behavior Rewarded?	96
	9.1.5 Are Extra Consequences Used Effectively?	97
	9.1.6 Is There a Skill Discrepancy?	97

		9.1.7 What Kind of Training Is Needed?	98
		9.1.8 Is the Person Right for the Job?	98
		9.1.9 In Summary	98
	9.2	Behavior-Based Safety Training	99
	9.3	Intervention and the Flow of Behavior Change	100
		9.3.1 Three Types of Behavior	101
		9.3.2 Three Kinds of Intervention Strategies	101
		9.3.3 The Flow of Behavior Change	102
		9.3.4 Accountability vs. Responsibility	103
	9.4	In Summary	103

PART FOUR Behavior-Based Intervention

Chapter 10 Intervening with Activators 107

	10.1	Principle #1: Specify the Desired Behavior	108
	10.2	Principle #2: Maintain Salience with Novelty	109
		10.2.1 Habituation	109
		10.2.2 Warning Beepers: A Common Work Example	110
	10.3	Principle #3: Vary the Message	110
		10.3.1 Changeable Signs	110
		10.3.2 Worker-Designed Safety Slogans	111
	10.4	Principle #4: Involve the Target Audience	111
		10.4.1 The "Flash for Life"	112
		10.4.2 The Airline Lifesaver	113
	10.5	Principle #5: Activate Close to Response Opportunity	114
	10.6	Principle #6: Implicate Consequences	114
		10.6.1 Incentives vs. Disincentives	115
		10.6.2 Setting Goals for Consequences	116
	10.7	In Summary	118

Chapter 11 Intervening with Consequences 119

	11.1	The Power of Consequences	119
		11.1.1 Intrinsic vs. Extrinsic Consequences	120
		11.1.2 Internal vs. External Consequences	122
	11.2	Managing Consequences for OSH	123
		11.2.1 The Case against Negative Consequences	123
		11.2.2 Discipline and Involvement	125
	11.3	"Dos" and "Don'ts" of Safety Rewards	127
		11.3.1 Doing It Wrong	127
		11.3.2 Doing It Right	127
		11.3.3 An Exemplary Incentive/Reward Program	130
		11.3.4 Safety Thank-You Cards	130
	11.4	In Summary	132

Chapter 12 Intervening as an Actively Caring Coach 133

	12.1	Intervening as a Safety Coach	133
		12.1.1 "C" for Care	134

Contents   xi

          12.1.2 "O" for Observe 136
          12.1.3 "A" for Analyze 141
          12.1.4 "C" for Communicate 141
          12.1.5 "H" for Help.. 144
   12.2 In Summary.. 145

Chapter 13 Intervening with Supportive Conversation 147
   13.1 The Power of Conversation 147
   13.2 The Art of Improving Conversation... 147
          13.2.1 Do Not Look Back 148
          13.2.2 Seek Commitment.. 148
          13.2.3 Stop and Listen 149
          13.2.4 Ask Questions First.. 150
          13.2.5 Transition from Nondirective to Directive 150
          13.2.6 Beware of Bias 151
   13.3 Recognizing Desirable Behavior 152
          13.3.1 Recognize During or Immediately after Desirable Behavior 153
          13.3.2 Make Recognition Personal for Both Parties 153
          13.3.3 Connect Specific Behavior with General
                    Higher-Level Praise 153
          13.3.4 Deliver Recognition Privately and One-on-One.................... 154
          13.3.5 Let Recognition Stand Alone and Soak In 154
          13.3.6 Use Tangibles for Symbolic Value Only................................ 154
          13.3.7 Special Advantages of Secondhand Recognition 154
   13.4 Receiving Recognition Well 155
          13.4.1 Avoid Denial and Disclaimer Statements.............................. 156
          13.4.2 Listen Attentively with Genuine Appreciation..................... 156
          13.4.3 Relive Recognition Later for Self-Motivation 156
          13.4.4 Show Sincere Appreciation 156
          13.4.5 Recognize the Person for Recognizing You 157
          13.4.6 Embrace the Reciprocity Principle... 157
          13.4.7 Ask for Recognition When Deserved but Not Forthcoming........ 157
   13.5 In Summary.. 157

PART FIVE *Actively Caring for People*

Chapter 14 Understanding Actively Caring 161
   14.1 What is Actively Caring?.. 161
          14.1.1 Three Ways to Actively Care 162
          14.1.2 Why Categorize AC4P Behaviors?... 163
          14.1.3 An Illustrative Anecdote 163
          14.1.4 A Hierarchy of Needs 165
   14.2 The Psychology of Actively Caring 166
          14.2.1 Lessons from Research 166
   14.3 A Consequence Analysis of AC4P Behavior.. 168
          14.3.1 The Power of Context 170
   14.4 Context at Work... 170
   14.5 In Summary.. 171

Chapter 15 The Person-Based Approach to Actively Caring 173
    15.1 Actively Caring from the Inside 173
        15.1.1 Person Traits vs. States 174
        15.1.2 AC4P States 174
    15.2 Actively Caring and Emotional Intelligence 179
        15.2.1 Safety, Emotions, and Impulse Control 180
        15.2.2 Nurturing EQ 181
    15.3 In Summary 181

Chapter 16 Increasing Occurrences of AC4P Behavior 182
    16.1 Enhancing the AC4P Person-States 182
        16.1.1 Self-Esteem 183
        16.1.2 Self-Efficacy 184
        16.1.3 Personal Control 185
        16.1.4 Optimism 187
        16.1.5 Belonging 187
    16.2 Directly Increasing Occurrences of AC4P Behavior 188
        16.2.1 Education and Training 189
        16.2.2 Consequences for Actively Caring 189
        16.2.3 The Reciprocity Principle 189
        16.2.4 Commitment and Consistency 190
    16.3 In Summary 192

PART SIX *Putting It All Together*

Chapter 17 Promoting High-Performance Teamwork 195
    17.1 Cultivating High-Performance Teamwork 195
        17.1.1 Selecting Team Members 195
        17.1.2 Clarify the Assignment 196
        17.1.3 Establish a Team Charter 197
        17.1.4 Develop an Action Plan 200
        17.1.5 Make it Happen 200
        17.1.6 Evaluate Team Performance 202
        17.1.7 Disband, Restructure, or Renew the Team 203
    17.2 In Summary 205

Chapter 18 Evaluating for Continuous Improvement 207
    18.1 Measuring the Right Stuff 207
    18.2 Developing a Comprehensive Evaluation Process 208
        18.2.1 What to Measure? 209
        18.2.2 Evaluating Environmental Conditions 209
        18.2.3 Evaluating Work Practices 211
        18.2.4 Evaluating Person Factors 211
        18.2.5 Evaluating Costs and Benefits 214
    18.3 You Can't Measure Everything 215
    18.4 In Summary 215

Contents xiii

Chapter 19 Obtaining and Maintaining Engagement 217
    19.1 Starting the Process 218
        19.1.1 Management Support 218
        19.1.2 Creating a Safety Steering Team 218
        19.1.3 Developing Evaluation Procedures 218
        19.1.4 Initiating an Education and Training Process 218
        19.1.5 Sustaining the Process 220
        19.1.6 Follow-Up Instruction/Booster Sessions 220
        19.1.7 Troubleshooting and Fine-Tuning 221
    19.2 Cultivating Continuous Support 221
        19.2.1 Where Are the Safety Leaders? 221
    19.3 Overcoming Resistance to Change 223
    19.4 Psychological Safety 227
        19.4.1 Relevance for OSH 228
        19.4.2 Cultivating Psychological Safety 228
        19.4.3 Great Leaders and Psychological Safety 229
    19.5 In Summary 230

Chapter 20 Reviewing the Principles 231
    20.1 Fifty Key Principles 231
        20.1.1 Principle 1: Safety Should Be Internally—
                Not Externally—Driven 231
        20.1.2 Principle 2: Culture Change Requires People to
                Understand the Principles and Know How to Use Them 232
        20.1.3 Principle 3: Champions of a TSC Will Emanate from
                Those Who Teach the Principles and Procedures 232
        20.1.4 Principle 4: Leadership Can Be Developed by Teaching
                and Demonstrating the Qualities of Effective Leaders 232
        20.1.5 Principle 5: Focus Recognition, Education, and Training
                on People Reluctant but Willing, Rather Than on Those
                Actively Resisting 232
        20.1.6 Principle 6: Giving People Opportunities for Choice Can
                Increase Commitment, Ownership, and Engagement 232
        20.1.7 Principle 7: A TSC Requires Continuous Attention to
                Factors in Three Domains: Environment, Behavior,
                and Person 233
        20.1.8 Principle 8: Do Not Count On Commonsense for Safety
                Improvement 233
        20.1.9 Principle 9: Safety Incentive/Reward Programs Should
                Focus on the Process Rather Than on Outcomes 233
        20.1.10 Principle 10: Safety Should Not be Considered a Priority
                but a Value with No Compromise 234
        20.1.11 Principle 11: Safety Is a Continuous Fight with Human
                Nature 234
        20.1.12 Principle 12: Behavior Is Learned from Three Basic
                Processes: Classical Conditioning, Operant Conditioning,
                and Observational Learning 234
        20.1.13 Principle 13: People View Behavior as Correct and
                Appropriate to the Degree They See Others Doing It 234

20.1.14	Principle 14: People Will Blindly Follow Authority, Even When the Mandate Runs Counter to Good Judgment and Social Responsibility	234
20.1.15	Principle 15: Group Participation Can Be Enhanced by Increasing Personal Responsibility, Individual Accountability, Group Cohesion, and Interdependence	234
20.1.16	Principle 16: On-the-Job Behavioral Observation and Interpersonal Behavioral Feedback Are Key to Achieving a TSC	235
20.1.17	Principle 17: Behavior-Based Safety Is a Continuous DO IT Process with D = Define Target Behaviors, O = Observe Target Behaviors, I = Intervene to Improve Behaviors, and T = Test Impact of the Intervention	235
20.1.18	Principle 18: Behavior Is Directed by Activators and Motivated by Consequences	236
20.1.19	Principle 19: Intervention Impact Is Influenced by the Amount of Response Information, Participation, and Social Support, as Well as External Consequences	236
20.1.20	Principle 20: Extra and External Consequences Should Not Over-Justify the Target Behavior	236
20.1.21	Principle 21: People Are Motivated to Maximize Positive Consequences (Rewards) and Minimize Negative Consequences (Costs)	236
20.1.22	Principle 22: Behavior Is Motivated by Six Categories of Consequences: Positive vs. Negative, Natural vs. Extra, and Internal vs. External	236
20.1.23	Principle 23: Negative Consequences Have Four Undesirable Side-Effects: Escape, Aggression, Apathy, and Countercontrol	237
20.1.24	Principle 24: Natural Variation in Behavior Can Lead to a Belief That Negative Consequences Have More Impact Than Positive Consequences	237
20.1.25	Principle 25: Long-Term Behavior Change Requires People to Change "Inside" as Well as "Outside"	237
20.1.26	Principle 26: All Perception Is Biased and Reflects Personal History, Prejudices, Motives, and Expectations	238
20.1.27	Principle 27: Perceived Risk Is Lowered When a Hazard Is Perceived as Familiar, Understood, Controllable, and Preventable	238
20.1.28	Principle 28: The Slogan "All Injuries Are Preventable" Is False and Reduces Perceived Risk	238
20.1.29	Principle 29: People Compensate for Increases in Perceived Safety by Taking More Risks	238
20.1.30	Principle 30: When People Evaluate At-Risk Behavior of Others, They Focus on Internal Factors; When Evaluating Their Own At-Risk Behavior, They Focus on External Factors	238
20.1.31	Principle 31: When Succeeding, People Over-Attribute Internal Factors, but When Failing, People Over-Attribute External Factors	240
20.1.32	Principle 32: People Feel More Personal Control and Perceived Choice When Working to Achieve Success Than When Working to Avoid Failure	240

	20.1.33	Principle 33: Stressors Lead to Positive Stress or Negative Distress Depending on Appraisal of Personal Control	240
	20.1.34	Principle 34: In a Total Safety Culture Everyone Goes Beyond the Call of Duty for OSH—They Actively Care for Safety	240
	20.1.35	Principle 35: Actively Caring Should Be Planned and Purposeful and Focus on Environment, Person, or Behavior	240
	20.1.36	Principle 36: Direct, Behavior-Focused AC4P Is Proactive and Most Challenging and Requires Effective Communication Skills	240
	20.1.37	Principle 37: Safety Coaching That Starts with Caring and Involves Observing, Analyzing, and Communicating Leads to Helping	241
	20.1.38	Principle 38: Actively Caring Can Be Increased Indirectly with Procedures That Enhance Self-Esteem, Belonging, and Empowerment	241
	20.1.39	Principle 39: Empowerment Is Facilitated with Increases in Self-Efficacy, Response-Efficacy, and Outcome Expectancy	241
	20.1.40	Principle 40: When People Feel Empowered, Their Safe Behavior Spreads to Other Situations and Behaviors	241
	20.1.41	Principle 41: AC4P Behavior Can Be Increased Directly by Educating People about Factors Contributing to Bystander Apathy	242
	20.1.42	Principle 42: As the Number of Observers of a Crisis Increases, the Probability of Helping Decreases	242
	20.1.43	Principle 43: AC4P Behavior Is Facilitated When Appreciated and Inhibited When Unappreciated	242
	20.1.44	Principle 44: A Positive Reaction to AC4P Behavior Can Increase the Benefactor's Self-Esteem, Empowerment, and Sense of Belonging	243
	20.1.45	Principle 45: The Universal Norms of Consistency and Reciprocity Motivate Everyday Behaviors, Including AC4P Behavior	243
	20.1.46	Principle 46: Once People Make a Commitment, They Encounter Internal and External Pressures to Think and Act Consistently with Their Position	243
	20.1.47	Principle 47: The Consistency Norm Is Responsible for the Impact of the "Foot-in-the-Door" Technique	244
	20.1.48	Principle 48: Managers Hold People Accountable; Leaders Inspire Self-Accountability	244
	20.1.49	Principle 49: When Numbers from Program Evaluations Are Meaningful to the Participants, They Can Direct and Motivate Intervention Improvement	244
	20.1.50	Principle 50: A Total Safety Culture Requires Psychological Safety	245
20.2	In Summary		245

Glossary of Key Terms ... 246
References ... 252
Index ... 255

Preface

Actively Caring for Safety teaches you the psychological science of occupational safety and health (OSH) in order to prevent the possibility of an unintentional injury to yourself and to others. Psychology influences every aspect of our lives, including our safety and health. The research-based principles of psychology or psychological science can be used to benefit almost every aspect of your life, including your safety and health. So, what is "psychology" anyway?

American Heritage Dictionary (1985) defines psychology as "1. the science of mental processes and behavior, 2. the emotional and behavioral characteristics of an individual, group, or activity" (p. 1000). Similarly, the two definitions in *The New Merriam-Webster Dictionary* (2023) are "1. the science of mind and behavior, 2. the mental and behavioral characteristics of an individual or group." In both dictionaries, the first definition of "psychology" uses the term "science" and refers to behavioral and mental processes.

Behaviors are the outside, objective, and observable aspects of people. Mental processes or mind reflects our inside, subjective, and unobservable characteristics—our cognitions, attitudes, and emotions. Science applies the scientific method for objective and systematic analysis and interpretation of reliable observations of natural and experimental phenomena.

So, what should you expect from a textbook on the psychological science of OSH? Obviously, such a book should reveal how psychology influences the safety and health of people. To be useful, such a text should explain and illustrate practical ways to apply psychological science in order to improve safety and health. That was, in fact, my purpose for creating this text—to teach you how to apply the science of human dynamics to both explain and reduce unintentional injuries to yourself and others.

As the science of human experience, psychology includes a vast number of subdisciplines. For example, areas covered in a standard college course in introductory psychology include research methods, physiological foundations, sensation and perception, language and thinking, consciousness and memory, learning, motivation and emotion, human development, intelligence, personality, psychological disorders, treatment of mental and behavioral disorders, social influence, environmental psychology and sustainability, industrial/organizational psychology, and human factors engineering.

The information in this book does not cover all of the domains of psychological science, only those directly relevant to understanding and influencing safety-related behaviors, cognitions, and attitudes. In addition, my coverage of information within any one sub-discipline of psychology is not comprehensive but focuses on only those features directly relevant to decreasing the occurrence of unintentional injuries in organizational and community settings.

This information will help you improve safety and health in any setting, from your home to the workplace and every community location in between. Actually, you can apply the knowledge gained from reading this text in all aspects of your daily life. However, most organized safety-improvement efforts occur in work environments, because that is where the exposure to hazardous conditions and at-risk behavior can be most devastating. As a result, most (but not all) of my illustrations and examples occur within an industrial or organizational context. However, you will readily see direct relevance of these principles and procedures to domains beyond the workplace.

For many years, the human side of OSH has been an extremely popular topic at national and regional safety and health conferences. Safety leaders realize that reducing injuries below current levels requires increased attention to the positive and negative contributions of people—the human factor. Engineering interventions and government policies have made their mark. Now it is time to include a focus on the human dynamics of safety—*the psychological science of injury prevention.*

A COMPREHENSIVE PERSPECTIVE

Most attempts to account for the human dynamics of safety have been limited in scope. Many trainers and consultants claim to have answers to the human side of safety, but their solutions are often impractical, shortsighted, or illusory. To promote/market their particular program, consultants, book authors, and conference speakers often give unfair and inaccurate criticism of alternative methods for addressing the human dynamics of injury prevention.

For example, behavior-based safety (BBS) has been criticized in an attempt to justify a focus on people's attitudes and/or cognitions. In contrast, promoters of BBS have ridiculed a focus on cognitive processes as being too subjective, unscientific, and unrealistic. Moreover, both behavior-based and attitude-focused approaches to injury prevention have been faulted in order to advocate for a systems or culture-focused approach. In fact, attitudes, behaviors, and cognitions require attention in order to develop large-scale and long-term improvements in people's safety and health; and proper attention to human dynamics (i.e., behaviors and person-states) enables the cultivation of an improved culture—ideally a Total Safety Culture (TSC).

Several books offer advice regarding the human side of OSH, but many of those texts provide a limited perspective. In fact, none of that scholarship is sufficiently comprehensive and practical to illustrate how to integrate behavior-based, cognitive-based, and attitude-based perspectives for a system-wide transformation to a TSC. My first textbook on the human dynamics of OSH, published in 1996 with the title *The Psychology of Safety: How to Improve Behaviors and Attitudes on the Job*, was created to do just that, and it was one of a kind at the time.

A slightly briefer paperback version of the hardcover textbook was also published in 1996 with the title *Working Safe: How to Help People Actively Care for Health and Safety*. In 2001, a second edition of each of those textbooks was published with updated and additional information. This text—*Actively Caring for Safety: The Psychological Science of Injury Prevention*—is a third edition of both 2001 textbooks on the human dynamics of cultivating and sustaining an injury-free workplace.

Simply put, principles from applied behavioral science and humanistic psychology—termed "humanistic behaviorism"—provide the basic tools and procedures for cultivating an effective injury-prevention OSH system, which requires total commitment and engagement from all potential participants. This text explains and illustrates practical ways to make that happen.

Recall that psychology includes the scientific study of both mind and behavior. Therefore, a practical book on the psychology of safety needs to teach science-based and practical approaches to improve what people think (cognition), feel (attitude), and do (behavior) in order to achieve a TSC.

CULTIVATING A TOTAL SAFETY CULTURE (TSC)

I refer to a TSC throughout this text as the ultimate vision of a safety-improvement mission. In a TSC, everyone feels responsible for safety and pursues it on a daily basis. At work, employees go beyond "the call of duty" to identify environmental hazards and at-risk behaviors, and whenever possible, they intervene to correct them. Safe work practices are supported with proper feedback/recognition procedures. In a TSC, safety is not a *priority* that gets shifted according to situational demands. Rather, safety is a *value* linked to all management directives and work priorities. Everyone in a TSC actively cares for safety.

Obviously, achieving a TSC requires a long-term continuous-improvement process. It involves the continuous improvement of OSH-related behaviors, cognitions, and attitudes of everyone in the culture. This book provides you with principles and procedures to do just that. Applying what you read here might not result in a TSC, but it will surely make a beneficial difference in your own safety and health and in the safety and health of others you choose to support.

I refer to helping others as "actively caring for people" (AC4P) that is reflected in the AC4P Movement my students and I initiated after the tragic shooting on our campus April 16, 2007

(Geller, 2008; www.gellerac4p.com). This book shows you how to improve the quality and increase the quantity of your own and others' AC4P behavior. Indeed, AC4P is the key to continuous improvement in safety and health. The more people who actively care for the safety and health of others, the less remote the achievement of our ultimate vision—a Total Safety Culture.

WHO SHOULD READ THIS BOOK?

Editors have warned me that one book can serve only a limited audience. However, a book that teaches practical strategies for reducing unintentional injuries on a large scale is relevant for everyone. All of us are at risk for an unintentional injury of some sort throughout the course of our days, and each of us can do something to reduce that risk to ourselves and to others. Therefore, a text that teaches practical ways to do that is pertinent reading for everyone.

However, the average person will not spend time reading a book on ways to reduce personal risk for unintentional injury. Why? Because most people don't believe they are at risk for personal injury. So why should they read a book about ways to improve safety? While I believe everyone *should* read this book, a text on the psychological science of safety is destined for a select and elite audience: people who are concerned about the rate of unintentional injuries in their organization or community and want to do something about it.

As indicated previously, this book is a third edition of two popular OSH textbooks, published first in 1996 and then in 2001, more than two decades ago. Every chapter in the second edition of my first OSH text was updated and expanded, and three new chapters were added—one on behavioral safety analysis, another on intervening with supportive conversation, and a third on promoting high-performance teamwork.

In addition to refined and updated written expression throughout this third edition, notable improvements include: a) language to replace common safety-related words that stifle human engagement (e.g., "accident investigation," "root cause," and "safety is a priority"); b) additional safety-relevant information on empathy, empowerment, emotional intelligence, risk compensation, self-motivation, and Maslow's Hierarchy of Needs; c) explicating the synergistic benefits of applying select principles of humanism with applied behavioral science (i.e., humanistic behaviorism); d) the dramatic benefits of promoting perceptions of personal choice; e) intermittent reference to relevant principles and procedures from positive psychology—a burgeoning domain of psychological science; f) critical distinctions between leadership and management as they relate to facilitating and supporting "psychological safety"—a relatively new term that includes many of the concepts and practices addressed in the earlier editions of this text; and g) a glossary of key terms used throughout this text.

Some language of psychologists is unique and quite different from that used in everyday discourse. Moreover, behavioral scientists commonly use terminology that is uncommon in other domains of psychological science. Plus, I have coined certain AC4P terms and defined some related psychological concepts in distinct ways. Thus, immediately before the references, a glossary of key terms is provided that explains several terms with a behavioral and/or AC4P focus, often resulting in a distinct difference from the more common usage of a certain term. You might benefit from referring to this glossary when coming across particular behavioral science or AC4P jargon in the description of a particular psychology principle or OSH process.

I am sensitive to the fact that a new edition of a textbook should justify its existence. I believe it is unfair to publish another edition of a book that is not a significant improvement over an earlier edition, although I have seen this happen too many times. Indeed, I have often purchased a follow-up edition of a book only to find very little difference between the two versions. This is often the case with college textbooks.

This book offers more research-based information than the 1996 and 2001 editions. Thus, readers of an earlier edition will not be disappointed if they purchase this revision. Plus, there are many potential applications of this text. It is a comprehensive source of psychological-science principles and practical procedures for the safety professional or corporate safety leader. It could be used as required or recommended reading in a number of undergraduate or graduate courses, especially courses on human factors engineering, safety management, organizational performance management, or industrial/organizational psychology.

Most engineering and psychology departments do not offer courses with safety or human factors in their titles. However, this text is quite suitable for such standard college/university courses as applied psychology, organizational psychology, management systems, engineering psychology, applied engineering, and even introductory psychology.

FUN TO READ

The writing style and format of this book are different than any professional text I have authored or read. Most authors of professional books, including me, have been taught a particular academic or research style of written expression that is not particularly enjoyable to read. When did you last pick up a nonfiction technical text for recreational or "fun" reading?

To attract more readership, this book is written in a more engaging style than most professional textbooks, thanks to invaluable editorial coaching for the prior editions by Dave Johnson, editor of *Industrial Safety and Hygiene News*. Moreover, each chapter includes several original drawings by George Wills to illustrate concepts and/or to add some humor to the teaching/learning process. I have interspersed some of those drawings in numerous professional keynote addresses and workshops, and audiences have found them both enjoyable and enlightening.

I predict some of you will page through this book and look for those illustrations. That is a useful beginning to learning concepts and techniques for improving the human dynamics of OSH. Later, you might read the explanation and applications of a particular psychological concept for a second useful step toward making a difference with this information. If you then discuss the principles and procedures with others, you will be on your way to applying this information in your organization, community, and/or your home.

TWO RELEVANT TESTIMONIES

Throughout this book, I include personal anecdotes to supplement the rationale of a principle or the description of an intervention technique or process. I end this preface with two anecdotes. The first testimony I reported in the prior editions of this text in order to set the stage for a memorable and enlightening learning experience. The second anecdote occurred after the second edition of this OSH book was published, and it reflects the continuous learning and improvement documented in this third edition of a text on the psychological science of OSH.

In August 1994, the Hercules Portland Plant stopped chemical production for two consecutive days so all 64 employees at the facility could receive my two-day workshop on the psychology of safety. Management had received a request for this all-employee workshop from a team of hourly workers who had previously attended my two-day professional development conference sponsored by the Mt. St. Helena Section of the American Society of Safety Engineers (ASSE).

Rick Moreno, a Hercules warehouse operator and hazardous materials unloader for more than 20 years, wrote the following reaction to my ASSE workshop. He actually read this to his coworkers at the start of the Hercules workshop. It set the stage for a most constructive and gratifying two days of education and training. If you approach the information in this book with some of the enthusiasm and optimism reflected in Rick's words, you will surely make a difference in someone's safety and/or health.

> Knowledge is precious. It's like trying to carry water in your cupped hands to a thirsty friend. Ideas that were crystal clear upon hearing them tend to slip from your memory like water through the creases of your hands, and while you may have brought back enough water to wet your friend's lips, he will not enjoy the full drink that you were able to take. And so it is with this analogy of a Total Safety Culture. Those who were there can only wet your lips with this new concept. Not a class nor

a program, but a safe way to live your life that spills into other avenues of our environment. It has no limit or boundaries as in this year, this plant. It's more like we're on our way and something wonderful is going to happen. And even though no answers are promised or given, the avenues in which to find our own answers for our own problems will be within our reach. That is why it is important that everyone has the opportunity to take a full drink of the Total Safety Culture instead of only having our lips wet. Something wonderful is going to happen.

This book is for you—Rick Moreno—and the many others who want to understand the psychological science of OSH and reduce the occurrence of unintentional injuries.

This second testimony came from a participant after he attended a half-day safety leadership workshop I had delivered in 2013. He shook my hand and said:

What a pleasure it was to hear your latest thoughts about person-to-person actively caring to benefit individuals, organizations, and communities. I first became aware of your research and scholarship when attending your day-long workshop at the ASSE convention in 2002. Since then I've read four of your books and taught my colleagues many of your principles.

That comment is not given to show off but rather to provide context for the rest of this individual's commentary.

After receiving that supportive feedback, I replied,

It's so nice to learn that my teachings are reaching others through the teaching of others. But since you've already read several of my recent books and taught others some of the contents, much of the workshop material I covered today was redundant, right?

The participant replied:

For sure, I understood where you were coming from, and I predicted where you were going throughout that session, and it was reassuring to hear it again. But what I really liked best was learning how your perspective, principles, and application suggestions have evolved over the ten years I've been following your work.

That last comment, stated a decade ago, reflects the critical point I want to make here. Practical ways to apply psychological science for solving real-world problems have progressed significantly over the years, through empirical research and numerous coaching/consulting experiences. It is so meaningful to have an organizational leader recognize, understand, and appreciate the evolution of recommended approaches for improving the complex human dynamics of organizational and societal challenges. Why? Because it justifies continuous collaboration and mutual learning from researchers and practitioners, and it validates the need to understand the rationale behind a process and to look continuously for ways to improve.

This third edition of two popular textbooks on the psychological science of OSH reflects my continuous learning from published research, personal experience, and the scholarship of many other safety professionals and practitioners. In many ways, my life has been one of never-ending researching, learning, and teaching in order to improve human welfare and life satisfaction on a large scale. It is my sincere hope that the information in this text will be used as a source of principles and procedures for guidance and benchmarks throughout various innovative journeys toward cultivating cultures of increasing numbers of people actively caring for the safety and health of others, as well as themselves.

In his classic bestseller—*The Seven Habits of Highly Effective People*—Stephen R. Covey wrote: "We live, we love, and we leave a legacy." This text on actively caring for safety is my legacy.

E. Scott Geller
December, 2023

Acknowledgments

In December 1992, I purchased an attractive print of a newborn colt from an artist at Galeria San Juan, Puerto Rico. While the artist—Jan D'Esopo—was signing my print, I asked her how long it took to complete the original. "Twenty-five minutes or 25 years," she replied, "depending on how you look at it." "What do you mean," I asked. "Well, it took me only 25 minutes to fill the canvas, but it took me 25 years of training and experience to prepare for the artistry."

I feel similarly about completing this book, which is a revision and expansion of the first edition published in 1996 and the second edition published in 2001. While writing the first and second editions and this revision took substantial time, that effort pales in comparison to the many years of preparation supported by invaluable contributions from teachers, researchers, consultants, safety professionals, university colleagues, and countless university students.

Actually, I have been preparing to write this text since entering the College of Wooster in Wooster, Ohio, in 1960. Almost all exams at that small liberal arts college required written discussion (rather than selecting an answer from a list of multiple-choice alternatives). Therefore, I received early experience and feedback at integrating concepts and research findings from a variety of sources. I was introduced to the scientific method at Wooster, and I applied it to my own psychological science research for an independent study project during both my junior and senior years.

Throughout five years of graduate education at Southern Illinois University in Carbondale, Illinois, I developed sincere respect and appreciation for the scientific method as the key to gaining profound knowledge. My primary areas of graduate study were learning, personality, social dynamics, and human information processing and decision-making. The Chair of both my thesis and dissertation committees (Dr. Gordon F. Pitz) gave me special coaching in research methodology and data analysis and helped me refine my skills for professional writing.

In 1968, I was introduced to the principles and procedures of applied behavior analysis (the foundation of behavior-based safety) from one graduate course and a few visits to Anna State Hospital in Anna, Illinois, where two eminent scholars—Drs. Ted Ayllon and Nate Azrin—were conducting seminal research in that research and scholarship domain. Those learning experiences (brief in comparison with all my other graduate-school education) convinced me that behavior-focused psychology could make large-scale improvements in people's lives. That insight had a dramatic impact on my future teaching, research, and scholarship.

I started my professional career in 1969 as assistant professor of psychology at Virginia Polytechnic Institute and State University (Virginia Tech). With assistance from undergraduate and graduate students, I developed a productive laboratory and research program in cognitive psychology. My tenure and promotion to associate professor was based entirely upon my professional scholarship in that domain. However, in the mid-1970s I became concerned that our laboratory work had limited potential for helping people. This conflicted with my personal mission to make beneficial large-scale differences in people's quality of life. Therefore, I turned to another line of research.

Given my conviction that applied behavioral science (ABS) has the greatest potential for solving organizational and community problems, I focused my research on finding ways to make that happen. Inspired by the first Earth Day in April 1970, my students and I developed, evaluated, and refined a number of community-based interventions for encouraging environment-constructive behavior and discouraging environment-destructive behavior. That prolific research program culminated with the 1982 Pergamon Press publication of *Preserving the Environment: New Strategies for Behavior Change*, which I co-authored with Drs. Richard A. Winett and the late Peter B. Everett.

Besides targeting environmental protection, my students and I applied behavioral science to a number of other problem areas, including prison administration, school discipline, community theft, transportation management, and alcohol-impaired driving. In the mid-1970s we began researching strategies for increasing the use of vehicle safety belts. That led to a focus on the application of behavioral science to prevent unintentional injuries in organizational and community settings.

Perhaps this brief history of my professional education and experience legitimizes my authorship of a book on the psychological science of safety. However, my purpose for providing that information was not to provide credibility but to acknowledge the vast number of individuals who prepared me to write this book. Critical for that preparation were our numerous research projects (since 1970), and that could not have been possible without dedicated contributions from hundreds of research students in our Center for Applied Behavior Systems (CABS), especially those diligent college seniors who coordinated the experiential training, data analysis, and scholarship of our research students. I am so very grateful for the dedication of all of those extraordinary leaders, but here I identify only the more recent coordinators of the research that influenced this third edition—Samuel Browning (2020–2021), Mackenzie Davis (2021–2022), Loralee Hoffer (2022–2023), and Tyler Parker-Rollins (2023–2024).

Financial support from a number of corporations and government agencies have made our 50 years of intervention research possible. Over the years, we received significant research funds from the Alcohol, Drug Abuse, and Mental Health Administration; the Alcoholic Beverage Medical Research Foundation; Anheuser-Busch Companies, Inc.; the Centers for Disease Control and Prevention; Domino's Pizza, Inc.; Exxon Chemical Company; General Motors Research Laboratories; the Motor Vehicle Manufacturers Association; the National Highway Traffic Safety Administration; the National Institute on Alcohol Abuse and Alcoholism; the National Institute for Occupational Safety and Health; the National Science Foundation; the U.S. Department of Education; the U.S. Department of Energy; the U.S. Department of Health, Education, and Welfare; the U.S. Department of Transportation; and the Virginia Departments of Agriculture and Commerce, Litter Control, Motor Vehicles, and Welfare and Institutions. Profound knowledge is only possible through programmatic research, and those organizations made it possible for my students and me to develop and evaluate various ways to improve attitudes and behaviors throughout organizations and communities.

I am also indebted to the numerous guiding and motivating communications I have received from corporate and community safety professionals worldwide. Daily contacts with those individuals shaped my research and scholarship and challenged me to improve connections between research and application. They also provided invaluable positive reinforcement to prevent "burnout." It would take pages to name all of those friends and acquaintances, and then I would necessarily miss many. You know who you are—thank you!

I need to acknowledge my informative and inspirational long-term collaboration with my colleagues at Safety Performance Solutions (SPS)—a leading-edge training and consulting firm that for more than four decades has taught organizations how to improve their safety culture and prevent personal injuries by applying the human dynamics revealed in this book. By working closely with the leaders of this group, especially Anne French, Mike Gilmore, Molly McClintock, Sherry Perdue, Chuck Pettinger, Steve Roberts, and Josh Williams, I have: a) developed meaningful empirical questions for programmatic research, b) learned how to make our research findings more applicable for OSH, and c) gained insight for teaching others the most relevant implications of our research. Specific contributions from SPS are mentioned in this text when relevant.

Vital to bridging the gap between research and application has been my long-term alliance and synergism with Dave Johnson, editor of *Industrial Safety and Hygiene News (ISHN)*. Dave and I began learning from each other in the spring of 1990 when I submitted my first article for his magazine. That year I submitted five articles on the psychology of safety, and Dave did

Acknowledgments

substantial editing for each one. Every time one of my articles was published, I learned something about communicating more effectively the bottomline of a psychological principle or procedure.

As an author of more than 350 research articles and former editor of the premier research journal in the applied behavioral sciences—*Journal of Applied Behavior Analysis*—I knew quite well how to write for a research audience in psychology. But Dave Johnson showed me that when it comes to writing for safety professionals and the general public, I had a lot to learn. Beginning in 1994, I authored a monthly article for a "Psychology of Safety" column in *Industrial Safety and Hygiene News* for 19 years. Each of those contributions profited immensely from Dave's suggestions and feedback. In fact, preparing those articles laid the groundwork for this book. Dave served as editor of the first edition of this text in 1996, dedicating long hours to improving the clarity and readability of my written expression. Thus, the talent and insight of Dave Johnson have been incorporated throughout this text, and I am eternally beholden to him.

The illustrations throughout this book were created by George Wills of Blacksburg, Virginia, which I think add vitality and fun to the written presentation. I am also sincerely grateful for the daily support and encouragement I have received from several of my PhD students—from 1993 until the present: especially Samuel Browning, Devin Carter, Tommy Cunningham, Jason DePasquale, Christopher Downing, the late Chris Dula, Kelli England, Trevin Glasgow, the late Kent Glindemann, Jeff Hickman, Angie Fournier, Phil Lehman, Timothy Ludwig, Chuck Pettinger, Steve Roberts, Doug Wiegand, Josh Williams, Ryan Smith, and Jack Wardale.

I am also indebted to the entire staff at CRC Press who brought this third edition to life, from editing and printing to distributing and marketing. Three individuals helped me most directly, beginning with James Hobbs—editor for CRC Press—who started me on the journey of preparing this third edition and has guided me along the way, and Kirsty Hardwick, who has advised my special assistant—Tyler Parker-Rollins—and me on preparing the text and illustrations for the publishing team, and Chris Mathews who led the process of refining the text for final printing and publication.

All of these people, plus many more, have contributed to my 60+ years of preparation to author *Actively Caring for Safety*. I thank you all very much. I am hopeful the synergy from all your contributions will help readers make rewarding and long-term positive differences in people's lives.

E. Scott Geller
December, 2023

About the Author

E. Scott Geller, Ph.D., an Alumni Distinguished Professor at Virginia Tech, is co-founder and senior partner of Safety Performance Solutions, Inc., a leading-edge training and consulting organization specializing in AC4P safety since 1995 (safetyperformance.com), and co-founder with his daughter, Krista Geller, PhD, of Geller AC4P, Inc., a consulting/training firm dedicated to teaching and spreading the Actively Caring for People (AC4P) Movement worldwide (gellerac4p.org). For 54 years, Professor Geller has taught and conducted research as a faculty member and Director of the Center for Applied Behavior Systems in the Department of Psychology at Virginia Tech.

He has authored, edited, or co-authored 54 books, 102 book chapters, 41 training manuals, 303 magazine articles, and more than 300 research articles addressing the development and evaluation of behavior-change interventions to improve quality of life on a large scale. His most recent textbook, with 31 coauthors: *Applied Psychology: Actively Caring for People* defines Dr. Geller's research, teaching, and scholarship career at Virginia Tech, which epitomizes the VT logo: *Ut Prosim*—"That I May Serve."

E. Scott Geller's dedication, competence, and energy helped him earn every university-wide teaching award offered at Virginia Tech. In 2001, Virginia Tech awarded him the University Alumni Award for Excellence in Research; in 2002, the university honored him with the Alumni Outreach Award for his exemplary real-world applications of behavioral science; and in 2003, he received the University Alumni Award for Graduate Student Advising. In 2005, he was awarded the statewide Virginia Outstanding Faculty Award by the State Council of Higher Education, and subsequently Virginia Tech honored him with the title of Alumni Distinguished Professor.

Dr. Geller is a fellow of the American Psychological Association, the Association for Psychological Science, the Association of Behavior Analysis International, and the World Academy of Productivity and Quality Sciences. He is past editor of the *Journal of Applied Behavior Analysis* (1989–1992); associate editor of *Environment and Behavior* (1982–2017); and current consulting editor for *Behavior and Social Issues*, the *Journal of Organizational Behavior Management*, and the *Journal of Safety Research*.

Dr. Geller has received lifetime achievement awards from the International Organizational Behavior Management Network (in 2008) and the American Psychological Foundation (in 2009). In 2010, he was honored with the Outstanding Applied Research Award from the American Psychological Association's Division of Applied Behavior Analysis, and in 2019, Professor Geller received the APA Division 25 Nathan H. Azrin Award for Distinguished Contributions to Applied Behavior Analysis. In 2024, The Virginia Association for Behavior Analysis initiated the annual E. Scott Geller Award for Distinguished Service to Applied Behavioral Science.

Part One

Orientation and Alignment

1 Choosing the Right Approach

The purpose of this book is outlined in this chapter—to explore the human dynamics of occupational safety and health (OSH) and to show how they can be managed to significantly improve safety-related performance. The principles and practical procedures you will learn are not based on commonsense nor intuition but rather on valid and reliable scientific investigation. Many recommendations seem counter to "pop psychology" and some traditional approaches to OSH. Therefore, please keep an open mind while you read about the psychological science of injury prevention.

> Organizations learn only through individuals who learn.
>
> (Peter Senge)

Safety professionals, team leaders, and concerned workers periodically scramble to find the "best" OSH technique for their workplace, especially after an injury or fatality. Typically, whatever offers the cheapest "quick fix" sells. This is not surprising, given the "lean and mean" atmosphere of the times. Programs that offer the most benefit with the least effort might sound best, but will they really work to improve OSH over the long term?

This text will help you ask the right questions to determine whether a particular approach to improving OSH will be effective. More importantly, this text describes the basic ingredients needed to improve organizational and community safety. In fact, you will find sufficient information to improve any OSH process. Learning the principles and procedures described here will enable you to make a beneficial, long-term difference in the safety and health of your workplace, home, and community. Actually, the information is relevant for most other performance domains—from increasing the quantity and quality of productivity in the workplace to improving quality of life in homes, neighborhoods, and throughout entire communities.

1.1 SELECTING THE BEST APPROACH

With so many different approaches available for improving OSH, how can a person select the best one? First, ask this question: "What does the research indicate?" In other words, are objective data available from program comparisons to inform the selection of a particular intervention? Unfortunately, there have been few systematic comparisons of alternative OSH interventions. However, this does not stop consultants from showing us impressive results regarding the success of their particular approach. Nor does it prevent them from implying (or boldly stating) that you can obtain similar fantastic results by simply following their patented "steps to success."

Keep in mind this marketing information usually comes from selected client case studies. Very few of those "success stories" were collected objectively and reliably enough to meet the rigorous standards of a professional research journal. When consultants try to "sell" you an approach for OSH with empirical data, ask them if they have published their results in a peer-reviewed journal. If they can show you a published research report of their impressive results or a professional presentation of a process very similar to theirs, then give their approach special consideration in your selection process. However, the validity and applicability of even published research varies dramatically. Figure 1.1 depicts the low end of research quality.

Most of the published research on safety improvement systematically evaluates whether a particular intervention worked in a particular situation, but it does not compare one OSH approach

FIGURE 1.1 Some research is not worth considering.

with another. In other words, such research tells us whether a particular strategy is better than nothing but offers no information regarding the relative impact of two or more different strategies for improving OSH. Such research has limited usefulness when selecting between different OSH interventions.

An exception can be found in a 1993 review article in *Safety Science*, where Stephen Guastello (1993) summarized systematically the evaluation data from 53 different research reports of OSH programs. Dr. Guastello provided rare and useful information for deciding how to improve OSH. You can assume the evaluations were both reliable and valid, because each report appeared in a scientific peer-reviewed journal. All of the studies selected for his summary were conducted in a workplace setting since 1977, and each study evaluated program impact with outcome data (including the number and severity of workplace injuries).

From my reading of Dr. Guastello's article, I believe it is safe to say the behavior-based and comprehensive ergonomics approaches lead the field. Personnel selection, the most popular method at the time (26 studies targeted a total of 19,177 employees), was among the least effective. Here are brief descriptions of those interventions implemented to reduce workplace injuries, ordered according to their effectiveness.

1.1.1 BEHAVIOR-BASED PROGRAMS

Programs in this category consisted of employee training regarding particular safe and at-risk behaviors, systematic observation and recording of the targeted behaviors, and feedback to workers regarding the frequency or percentage of safe vs. at-risk behavior. Some of these programs included goal-setting and/or incentives to encourage the observation-and-feedback process.

1.1.2 COMPREHENSIVE ERGONOMICS

The ergonomics (or human factors) approach to OSH refers essentially to any adjustment of working conditions or equipment in order to reduce the frequency or probability of an environmental hazard or at-risk behavior. An essential ingredient in those programs was a diagnostic survey or environmental audit by employees, leading to specific recommendations for eliminating hazards that put employees at-risk or promoted at-risk behaviors.

1.1.3 Engineering Changes

This category includes the introduction of robots or the comprehensive redesign of facilities to eliminate certain at-risk behaviors. It is noted, however, that the robotic interventions introduced the potential for new types of workplace injuries, such as a robot catching an operator in its work envelope and impaling him or her against a structure. Thus, robotic innovations usually require additional engineering intervention such as equipment guards, emergency kill switches, radar-type sensors, and workplace redesign to prevent injury from robots. Behavior-based training, observation, and feedback (as detailed in Section 4 of this book) are also needed following engineering redesign.

1.1.4 Group Problem Solving

For this approach, operations personnel met voluntarily to discuss safety issues and problems and to develop action plans for safety improvement. This approach is analogous to quality circles where employees who perform similar types of work meet regularly to solve problems of product quality, productivity, and cost.

1.1.5 Government Action (in Finland)

In Finland, two government agencies responsible for labor production targeted the most problematic occupational groups and implemented certain action strategies. These included:

- Disseminating information to work supervisors regarding the cause of workplace injuries and methods to reduce them.
- Setting standards for safe machine repair and use.
- Conducting periodic work-site inspections.

1.1.6 Management Audits

For the programs in this category, designated managers were trained to administer a standard International Safety Rating System (ISRS). This system evaluates workplaces based on 20 components of OSH, including leadership and administration, management training, planned inspections, task and procedural analysis, accident investigations, task observations, emergency preparedness, organizational rules, accident analysis, employee training, personal protective equipment, health control, program evaluation, engineering controls, and off-the-job safety.

Managers conduct the comprehensive audits annually to develop improvement strategies for the next year. Specially certified ISRS personnel visit target sites and recognize a plant with up to five "stars" for exemplary safety performance.

1.1.7 Stress Management

These programs taught employees how to cope with stressors or the sources of work stress. Exercise was often a key action strategy promoted as a way to prevent stress-related injuries in physically demanding jobs. The topic of stress as it relates to injury prevention is presented in Chapter 6.

1.1.8 POSTER CAMPAIGNS

The two published studies in this category evaluated the injury-reduction impact of posting signs that urged workers at a shipyard to avoid certain at-risk behaviors and to follow certain safe behaviors. Most signs were posted at relevant locations and gave specific behavioral instructions like: "Take material for only one workday," "Gather hoses immediately after use," "Wear your safety helmet," and "Check railing and platform couplings (on scaffolds)."

For one study, safety personnel at the shipyard gave work teams weekly feedback regarding compliance with sign instructions. In the other study, environmental audits, group discussions, and structured interviews were used to develop the poster messages. Thus, it is possible that factors other than the posters themselves contributed to the moderate short-term impact of this intervention approach. All of these factors are covered in this text, including ways to maximize the beneficial effects of safety signs (Chapter 10).

1.1.9 PERSONNEL SELECTION

This popular but relatively ineffective approach to injury prevention is based on the intuitive notion of "accident proneness." The strategy is to identify aspects of accident proneness among job applicants and then screen out people with critical levels of certain characteristics.

Although measuring and screening for accident proneness sounds like a "quick fix" approach to injury prevention, this method has several problems you will readily realize as you read more in this book about the psychological science of OSH. Briefly, this technique has not worked reliably to prevent workplace injuries because:

- The instruments or procedures available to measure the proneness characteristics are unreliable or invalid.
- The characteristics do not generalize across settings, so a person might show them at home but not at work, or vice versa.
- A person with a higher desire to take risks (such as a sensation seeker) might be more inclined to take appropriate precautions (like using personal protective equipment) to avoid potential injury.

1.1.10 NEAR MISS* REPORTING

This approach involved increased reporting and investigation of incidents that did not result in an injury but certainly could have under slightly different circumstances. One program in this category increased the number of corrective suggestions generated but did not reduce injury rate. The other scientific publication in this category reported a 56 percent reduction in injury severity as a result of increased reporting of close calls, but the overall number of injuries did not change.

1.2 THE CRITICAL HUMAN ELEMENT

Every OSH approach described previously requires that you consider the human element or the psychology of safety. Indeed, the most successful approaches, behavior-based safety and comprehensive ergonomics, directly address the human aspects of OSH. The bottom line is illustrated in Figure 1.2. The three employees are looking at a contributing factor in almost every injury—the human factor. Thus, any safety intervention that improves the safety-related behaviors of workers will prevent workplace injuries.

The behavior-based approach targets human behavior and relies on interpersonal observation and feedback for intervention. The success of comprehensive ergonomics depends on employees observing relationships between behaviors and work situations and then recommending

Choosing the Right Approach

FIGURE 1.2 Human dynamics contribute to almost every injury.

feasible changes in behavior, equipment, or environmental conditions to make the job more "user friendly" and safe.*

Obviously, achieving success in OSH requires concerted efforts in the domain of psychological science. Ever since the first edition of this text was published in 1996, seminars at national and regional safety conferences that purported to teach aspects of the psychology of safety have consistently attracted large audiences. Consider, for example, the following titles from conferences of the National Safety Council or the American Society of Safety Engineers.

- "Managing Safe Behavior for Lasting Change"
- "Humanizing the Total Safety Program"
- "The Human Element in Achieving a Total Safety Culture"
- "The Psychology of Injury Prevention"
- "Behavior-Based Safety Management: Parallels with the Quality Process"
- "Behavioral Management Techniques for Continuous Improvement"
- "Improving Safety Through Innovative Behavioral and Cultural Approaches"
- "Safety Leadership Power: How to Empower All Employees"
- "Moving to the Second Generation in Behavior-Based Safety"
- "Potholes in the Road to Behavioral Safety"
- "Implementing Behavior-Based Safety on a Large Scale"
- "Motivating Employees for Safety Success"
- "Integrating Behavioral Safety into Other Safety Management Systems"
- "From Knowing to Doing: Achieving Safety Excellence"
- "Safety and Psychology: Where Do We Go from Here?"

---

* "Near miss" is used routinely in the workplace to refer to an incident that did not result in an injury. Since a literal translation of this term means the injury actually occurred, "close call" is used throughout this book instead of "near miss."

I attended each of those presentations and found numerous inconsistencies between presentations dealing with the same topic. Sometimes, I noted erroneous and frivolous statements, inaccurate or incomplete references to psychological theory or research, and invalid or irresponsible comparisons between various approaches to dealing with the psychology of safety. It seemed a primary aim of several presentations was to "sell" their own particular program or consulting services by overstating the benefits of their approach and giving an incomplete or naive discussion of alternative methods or procedures. None of those presentations referred to a research publication for evidence of effectiveness.

1.3 THE FOLLY OF CHOOSING WHAT SOUNDS GOOD

The theory, research, and intervention techniques from psychological science are so vast and often so complex that it can be overwhelming to decide which particular approach or strategy to use. As a result, we are easily biased by commonsense words that sound good. As depicted in Figure 1.3, commonsense is subjective, based on a person's everyday *selective* experiences and biased interpretations of those experiences.

Valid theory, principles, and procedures founded on empirical research evidence are often ignored. In fact, an apparently endless market of self-help books, audiotapes, and videotapes address concepts seemingly relevant to the psychology of injury prevention. For example, I have listened to the following audiotapes that represent only a fraction of the "pop psychology" presentations with topics relevant to the psychology of safety.

- "Coping with Difficult People" by R. M. Branson
- "Personal Excellence" by K. Blanchard
- "How to Build High Self-Esteem" by J. Canfield
- "The Seven Habits of Highly Effective People" by S. R. Covey
- "First Things First" by S. R. Covey, A. R. Merrill, and R. R. Merrill
- "The Science of Personal Achievement" by N. Hill

FIGURE 1.3 Without science, decision-making is a biased shot in the dark.

Choosing the Right Approach

- "Increasing Human Effectiveness" by R. Moawad
- "Lead the Field" by E. Nightingale
- "Unlimited Power" by A. Robbins
- "The Psychology of Achievement" by B. Tracey
- "The Psychology of Success" by B. Tracey
- "The Universal Laws of Success and Achievement" by B. Tracy
- "The Psychology of Winning" by D. Waitley
- "Self-Esteem" by J. White
- "Goal Setting" by Z. Ziglar
- "Top Performance" by Z. Ziglar
- "The Secrets of Power Persuasion" by R. Dawson
- "The 12 Life Secrets" by R. Stuberg
- "The Courage to Live Your Dreams" by L. Brown
- "The New Dynamics of Goal Setting" by D. Waitley
- "Transforming Stress into Power" by M. J. Tazer and S. Willard

Which, if any, of those pop psychology audiotapes gives safety professionals the "truth"—the most effective and practical tools for dealing with the human dynamics of OSH? Some of the most cost-effective strategies for managing behaviors and attitudes at the personal and organizational level are not even mentioned in many of the pop psychology books, audiotapes, and videotapes. That might be the case not only because authors and presenters are unaware of the latest research but also because many of the best techniques for individual and group improvement do not sound good—at least at first. The primary purpose of this text is to teach the most effective approaches for addressing and improving the human dynamics of OSH. The principles and procedures were not selected because they sound good, but because their validity has been supported with valid research.

1.4 RELYING ON RESEARCH

This book teaches research-based psychology related to OSH. Thus, by reading this text, you will improve your commonsense about the human dynamics of OSH. At this point, I hope you are open to questioning the validity of good-sounding statements that are not supported by quality research. Psychological science, for example, does not generally support the following common statements related to the human dynamics of OSH:

- Practice makes perfect.
- Spare the rod and spoil the child.
- Attitudes need to be changed before behavior will change.
- Human nature motivates safe and healthy behavior.
- People will naturally help in a crisis.
- Rewards for not having injuries reduce injuries.
- All injuries are preventable.
- Zero injuries should be a safety goal.
- Manage only that which can be measured.
- Safety should be considered a priority.

These and other common safety messages and apparent beliefs will be refuted in this book, with reference to scientific knowledge obtained from systematic research. Sometimes, case studies will illustrate the practicality and benefits of a particular principle or procedure, but the validity of the information was not founded on case studies alone. The approaches presented in this text were originally discovered and verified with empirical and repeated scientific research in laboratory and field settings.

1.5 START WITH BEHAVIOR

Many pop psychology self-help books, audiotapes, and motivational speeches give minimal if any attention to behavior-based approaches to personal achievement. "Behavioral control" and "behavior modification" do not sound good. Even the term "behavior" has negative connotations, as in "let's talk about your behavior at the party last night." B. F. Skinner, the founder of behavioral science and its many practical applications, was one of the most misunderstood and underappreciated scientists and scholars of the 20th century, primarily because the behavior management principles he taught did not sound good.

Professor Skinner and his followers have shown repeatedly that behavior is motivated by its consequences, and thus behavior can be changed by controlling the events that follow behavior. However, this principle of "control by consequences" does not sound as good as "control by positive thinking and free will." Therefore, the research-based principles and procedures from behavioral science have been underappreciated and underused.

This book teaches you how to apply behavioral science for OSH. The research recommends we start with behavior. However, the demonstrated validity of a behavior-based approach does not mean the better-sounding, person-based approaches should not be used. It is important to consider the feelings and attitudes of employees, because these individuals are needed to implement the tools of behavior management.

This text will teach you how certain person-states critical for preventing an unintentional injury—self-esteem, empowerment, and self-motivation—can be increased by applying principles from behavioral science and humanism (i.e., humanistic behaviorism). It is possible to establish interpersonal interactions and behavioral consequences in the workplace to increase positive OSH-related feelings and attitudes. You will learn how increasing those person-states benefits behavior and helps to cultivate and achieve a Total Safety Culture (TSC).

As illustrated in Figure 1.4, an attitude of frustration or an internal person-state of distress can certainly influence driving behavior, and vice versa. Indeed, internal (unobserved) states of mind continually influence observable behaviors, while changes in observable behaviors continually affect changes in person-states or attitudes. Thus, it is possible to "think a person into safe

FIGURE 1.4 Behavior influences attitude, and attitude influences behavior.

Choosing the Right Approach

behavior" (through education, coaching, and consensus-building exercises), and it is possible to "act a person into safe thinking" (through humanistic behaviorism).

In an industrial setting, it is most cost effective to target behaviors first through behavior management interventions (described in this text) implemented by employees themselves. Small changes in behavior can result in attitude change, followed by more behavior change and more desired attitude change. This spiraling of behavior feeding attitude, attitude feeding behavior, behaviors feeding attitudes, and so on can lead to employees becoming totally committed to OSH achievement, as reflected in their daily behavior. All of this could start with relatively insignificant behavior change in one employee—a "small win."

1.6 IN SUMMARY

In this initial chapter, I outlined the purpose of this text, which is to teach principles for understanding the human dynamics of OSH and to illustrate practical evidence-based procedures for applying these principles to achieve significant improvements in organizational and community-wide safety and health.

The principles and procedures are not based on commonsense or intuition but rather on reliable scientific investigation. Some of these will contradict common folklore or pop psychology and require shifts in traditional approaches to the management of OSH. Approach this material with an open mind. Be prepared to relinquish fads, fancies, and folklore for innovations based on unpopular but research-supported theories and principles.

I promise the advice herein for "Actively Caring for Safety" is based on the latest and most reliable scientific evidence related to individual and organizational safety and health. As illustrated in Figure 1.5, "trying harder" cannot replace proper equipment, method, and tools. This book provides research-tested methods and tools for improving the human dynamics of OSH and thereby preventing injuries at work, at home, and throughout your community.

FIGURE 1.5 Motivation can't replace equipment and method.

2 Starting with Theory

In this chapter we consider the value of theory in guiding our approaches to improving OSH. You will see how a vision for a Total Safety Culture (TSC) is a useful guide to achieve OSH excellence. A basic principle here is that effective OSH performance results from the dynamic interaction of environment, behavior, and person-based factors. Achieving a TSC requires attention to each of these domains. A case is made for integrating person-based and behavior-based psychology (i.e., humanistic behaviorism) in order to address most effectively the human dynamics of injury prevention.

> There's nothing so practical as a good theory.
>
> **(Kurt Lewin)**

As you know, some OSH efforts suffer from the "flavor of the month" syndrome. New procedures or intervention programs are tried seemingly at random, without an apparent vision, plan, or supporting set of principles. When the mission and principles are not clear, employees' acceptance and involvement suffer.

A theory or set of guiding principles makes it possible to evaluate the consistency and validity of program goals and intervention strategies. By summarizing the appropriate theory or principles into a mission statement, you have a standard for judging the value of your company's procedures, policies, and performance expectations.

It is important to develop a set of comprehensive principles on which to base your OSH procedures and policies. Then, teach those principles to your employees so they are understood, accepted, and appreciated. Such buy-in is certainly strengthened when employees or associates help select the OSH principles to follow and summarize them in a company mission statement.

This is theory-based OSH. A critical challenge, of course, is to choose the most relevant theories or principles for your company culture and purpose, and to develop an appropriate and feasible mission statement that reflects the right theory.

2.1 THE MISSION STATEMENT

Several years ago, I worked with employees of a major chemical company to develop the general mission statement for OSH given in Figure 2.1. This vision for a Total Safety Culture serves as a guideline or standard for the material presented throughout this book, in the same way that a corporate mission statement provides a yardstick for gauging the development and implementation of various policies and procedures.

That OSH mission statement might not be suitable for all organizations, but it is based on appropriate and comprehensive theory, supported by empirical data from research in psychology. Before developing this statement, employees learned basic theories of psychological science most relevant to improving OSH. Those principles are illustrated throughout this text, along with operational (real-world) definitions.

2.2 THEORY AS A MAP

The following personal experience exemplifies how a theory can be viewed as a map to guide us to a destination. The mission statement in Figure 2.1 reflects a destination for OSH within the realm of psychological science. This story also reflects the difficulty in finding the best theory among numerous possibilities.

> **Mission Statement**
>
> A Total Safety Culture (TSC) continually improves safety performance. To that end, a TSC:
>
> Promotes a work environment based on employee involvement, ownership, team work, education, training, and leadership.
>
> Builds self–esteem, empowerment, pride, enthusiasm, optimism, and encourages innovation.
>
> Reinforces the need for employees to actively care about their fellow coworkers.
>
> Promotes the philosophy that safety is not a priority that can be reordered, but is a value associated with every priority.
>
> Recognizes group and individual achievement.

FIGURE 2.1 The principles and procedures covered in this book are reflected in this safety-achievement mission statement.

I had the opportunity to conduct a training program at a company in Palatka, Florida. My client sent me step-by-step instructions to take me from Interstate 95 to Palatka. That was my map, limited in scope for sure, but presumably sufficient to get the job done—to get me to the Holiday Inn in Palatka. However, while at the National Car Rental desk, an attendant said my client's directions were incorrect and showed me the "correct way" with National's map of Jacksonville.

I was quick to give up my earlier theory (from my client's handwritten instructions) for this more professional display. After all, I now had a professionally printed map and directions from someone in the business of helping customers with travel plans—a consultant, so to speak. But National's map showed details for a limited area, and Palatka was not on the map. I could not verify the attendant's directions with the map, nor could I compare those directions with my client's very different instructions. Without a complete perspective, I chose the theory that looked best, and I got lost. As depicted in Figure 2.2, it is critical to start out with the right map (or theory).

After traveling 15 miles, I began to question the "National theory" and wondered whether my client's scribbling had been correct after all. Nevertheless, I drove another ten miles before exiting the highway in search of further instructions. I certainly needed to reach my destination that night, but motivation without appropriate direction can do more harm than good. In other words, a motivated worker cannot reach safety goals with the wrong theory or principles.

It was late Sunday night and the gas station off the exit ramp was closed, but another vehicle had also just stopped in the parking lot. I drove closer and announced to four tough-looking, grubby characters in a pick-up truck loaded with motorcycles that I was lost and wondered if they knew how to get to Palatka. None of those men had heard of Palatka, but one pulled out a detailed map of Florida and eventually found the town of Palatka. I could not see the details in the dark, but I accepted this new "theory" anyway, with no personal verification.

The packaging of this theory was not impressive, but my back was against the wall. I was desperate for a solution to my problem, and I had no other place to turn. As I left the parking lot with a new theory, I wondered whether I was now on the right track. Perhaps, the theory obtained from the National Car rental attendants was correct and I had missed an exit. Who should I trust? Fortunately, I looked beyond the slick packaging and went with the guy who had the more comprehensive perspective (the larger map). That theory got me to the Holiday Inn Palatka.

FIGURE 2.2 Start your journey with the right theory.

2.3 RELEVANCE TO OSH

That evening I thought about my experience and its relevance to OSH. It reminded me of the dilemma facing many safety professionals when they choose approaches, programs, and consultants to help solve people issues related to OSH. Theories, research, and tools in psychology are so vast and often so complex that it can be an overwhelming task to select a theory or a set of principles to follow.

The theory that got me to Palatka was not the most professional or believable, nor was it "packaged" impressively. This does not mean you should avoid the slick, well-marketed approaches to OSH. I only wish such factors were given much less weight than empirical data.

However, it is relevant that the more comprehensive map enabled me to find my destination. Relatedly, I have found that many of the approaches to improving the human dynamics of OSH are limited in scope or theoretical foundation. Many are sold or taught as packaged programs or step-by-step procedures for any workplace culture.

In the long run, it is more useful to teach relevant theory and principles. On a solid research-supported foundation, OSH procedures and interventions can be customized by employees who will "own" and follow them. As the old saying goes, "Give a man a fish and you feed him for a day, teach him how to fish and you feed him for a lifetime."

At breakfast, I told the human resource manager and the safety director—the one who gave me the handwritten instructions—about my problems finding Palatka. Interestingly, each had a different theory on the best way to travel between the Jacksonville airport and Palatka. The safety director stuck with his initial instructions, whereas the human resource manager recommended the route I eventually took. Their discussion was not enlightening. In fact, it got me more confused, because I did not have a visual picture or schema (a comprehensive map) in which to fit the various approaches (or routes) they were discussing. In other words, I did not have a framework or paradigm to organize their verbal descriptions. Without a relevant theory my experience taught me nothing, except the need for an appropriate and understandable theory—in this case a map.

A theory for injury prevention should serve as the map that provides direction to meet a specific OSH challenge. Obviously, it is important to teach the basic theory to everyone who must meet the challenge. Then it is a good idea to have an employee task force summarize the theory in a safety mission statement. When the workforce understands the theory and accepts the summary mission statement, intervention processes based on that theory will not be viewed as "flavor of the month" but as an action plan to bring the theory to life.

When employees appreciate and affirm the theory, they will get involved in designing and implementing the various action steps. They will also suggest ways to refine or expand action plans and theory on the basis of systematic observations or scientific evidence. This is the best kind of continuous improvement.

2.4 A BASIC MISSION AND THEORY

The mission statement in Figure 2.1 reflects the ultimate vision for OSH—a Total Safety Culture. In a TSC:

- Everyone feels responsible for safety and does something about it on a daily basis.
- Participants go beyond the call of duty to identify unsafe conditions and at-risk behaviors, and they intervene to correct them.
- Safe work practices are encouraged intermittently with supportive feedback from both peers and managers.
- People "actively care" routinely for the safety of themselves and others.
- Safety is not considered a priority that can be conveniently shifted depending on the demands of the situation; rather, safety is considered a core value linked to every priority of a given situation.

That TSC vision is much easier said than done, but it is achievable through a variety of safety processes rooted in the disciplines of engineering and psychological science. Generally, a TSC requires continual attention to three domains.

1. Environment-based factors (including equipment, tools, physical layout, procedures, standards, and temperature).
2. Person-based factors (including people's attitudes, beliefs, and dispositions).
3. Behavior-based factors (including safe and at-risk work practices, as well as going beyond the call of duty to intervene on behalf of another person's safety).

This triangle of safety-related factors is illustrated in Figure 2.3. The various factors are dynamic and interactive. A change in one domain eventually impacts the other two. For example,

FIGURE 2.3 A Total Safety Culture requires continual attention to three domains of contributing factors.

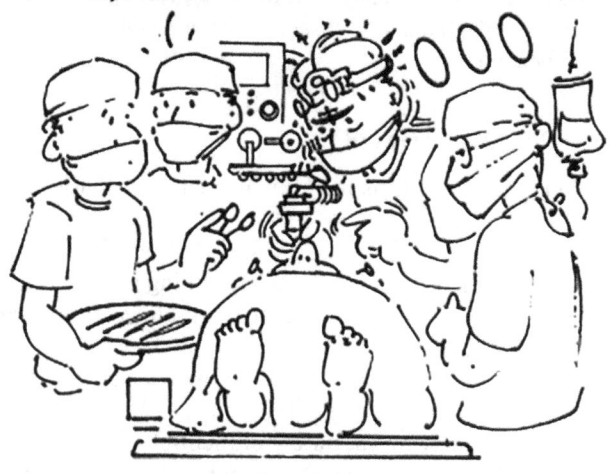

FIGURE 2.4 Performance results from the dynamic interaction of environment, behavior, and person factors.

behaviors that reduce the probability of injury often involve environmental change and lead to attitudes consistent with the safe behaviors. The behavior-based and person-based factors represent the human dynamics of OSH and are addressed in this book.

Paying attention to only behavior-based factors (the observable activities of people) or to only person-based factors (the unobservable feeling states or attitudes of people) is like using a limited map to find a destination, as with my attempt to find Palatka, Florida. The mission to achieve a TSC requires a comprehensive framework—a complete map of the relevant psychological territory. Figure 2.4 illustrates the complex interaction of environment, behavior, and person-states.

2.5 BEHAVIOR-BASED VS. PERSON-BASED APPROACHES

Psychological science has influenced and informed a variety of interventions to benefit people and organizations. Most of these can be classified into one of two basic approaches: person based and behavior based. In fact, most of the numerous psychotherapies available to treat developmental disabilities and psychological disorders—from neurosis to psychosis—can be classified as essentially person-based or behavior-based.

More specifically, most psychotherapies focus on changing people either from the inside ("thinking people into acting differently") or from the outside ("acting people into thinking differently"). Person-based approaches attack individual attitudes or thinking processes directly. They teach clients new thinking strategies or give them insight into the origin of their abnormal or unhealthy thoughts, attitudes, or feelings. In contrast, behavior-based approaches attack a client's behaviors directly. They change relationships between behaviors and their consequences. This text will show you how to integrate relevant principles from these two psychological approaches in order to achieve a TSC.

2.6 THE PERSON-BASED APPROACH

Imagine you see two employees pushing each other in a parking lot as a crowd gathers around to watch. Is this aggressive behavior, horseplay, or mutual instruction for self-defense? Are the employees physically attacking each other to inflict harm, or does this physical contact indicate a special friendship and mutual understanding of the line between aggression and play? Perhaps if you watch longer and pay attention to verbal behavior, you will decide whether this is aggression, horseplay, or a teaching/learning demonstration.

However, a truly accurate account might require you to assess each individual's personal feelings, attitudes, or intentions. It is possible, in fact, that one person was being hostile while the other was just having fun, or the contact started as horseplay and progressed to aggression.

This scenario illustrates a basic premise of the person-based approach. Focusing only on observable behavior does not explain enough. People are much more than their behavior. Concepts like intention, creativity, intrinsic motivation, subjective interpretation, self-esteem, and mental attitude are essential to understanding and appreciating the human dynamics of a problem. Thus, a person-based approach in the workplace applies surveys, personal interviews, and focus-group discussions to find out how individuals feel about certain situations, conditions, behaviors, or personal interactions.

Humanism is the most popular person-based approach today, as evidenced by the current market of pop psychology videotapes, audiotapes, and self-help books. The key principles of humanism found in most pop psychology approaches to increase personal achievement are:

1. Everyone is unique in numerous ways. The special characteristics of individuals cannot be understood or appreciated by applying general principles or concepts, such as the behavior-based principles of performance management or the inborn personality-trait perspective of psychoanalysis.
2. Individuals have far more potential to achieve than they typically realize, and they should not feel hampered by past experiences or present liabilities.
3. The present state of an individual in terms of feeling, thinking, and believing is a critical determinant of personal success.
4. One's self-concept influences mental and physical health, as well as personal effectiveness and achievement.
5. Ineffectiveness and abnormal thinking and behavior result from large discrepancies between one's real self ("who I am") and one's ideal self ("who I would like to be").
6. Individual motives vary widely and come from within a person.

2.7 THE BEHAVIOR-BASED APPROACH

The behavior-based approach to applied psychology is founded on behavioral science as conceptualized and researched by B. F. Skinner (1938, 1974). In his experimental analysis of behavior, Professor Skinner rejected for scientific study unobservable factors such as self-esteem, intentions, and attitudes. He researched only observable behavior and its social, environmental, and physiological determinants. The behavior-based approach starts by identifying observable behaviors targeted for change and the environmental conditions that can be manipulated to influence the target behavior(s) in desired directions.

The critical premise is that behavior can be studied objectively and changed by identifying and manipulating environmental conditions (or stimuli) that immediately precede and follow a target behavior. The antecedent conditions (which I call "activators") signal when behavior can achieve a pleasant consequence (a reward) or avoid an unpleasant consequence (a penalty). Therefore, activators direct behavior, and consequences determine whether the behavior will recur. Accordingly, people are motivated by the consequences they expect to receive, escape, or avoid after performing a target behavior.

Humanists maintain that this activator-behavior-consequence (ABC) analysis is much too simplistic to explain human behavior. For many applications, they are right. However, as shown in Figure 2.5, many of our daily behaviors are directed by preceding activators and motivated by ensuing consequences. Much more about this ABC approach to understanding and improving

FIGURE 2.5 It is an S-R world after all.

behavior is provided in Parts 3 and 4 of this book. Then in Part 5, the person-based approach of the humanists is explained, especially regarding a synergistic integration with the behavior-based approach (i.e., humanistic behaviorism) to bring out the best in people and their organizations for the purpose of achieving a TSC.

2.8 CONSIDERING COST EFFECTIVENESS

When people act in certain ways, they usually adjust their mental attitude and self-talk to parallel their actions; when people change their attitudes, values, or thinking strategies, certain behaviors change as a result. Thus, person-based and behavior-based approaches to improving people can influence both attitudes and behaviors, either directly or indirectly. Most parents, teachers, first-line supervisors, and safety managers use both approaches in their attempts to benefit a person's knowledge, skills, attitudes, and/or behaviors.

- When we lecture, counsel, or educate others in a one-on-one or group situation, we are essentially using a person-based approach.
- When we recognize, correct, or discipline others for what they have done, we are operating from a behavior-based perspective.

Unfortunately, we are not always effective with our person-based or behavior-based intervention techniques, and often we do not know whether our intervention worked as intended. In order to apply person-based techniques to psychotherapy, clinical psychologists receive specialized therapy or counseling training for four or more years, followed by an internship of at least one year. This intensive training is needed because tapping into an individual's perceptions, attitudes, and thinking styles is a demanding and complex process. Also, the internal dimensions of people are extremely difficult to measure reliably, making it cumbersome to assess therapeutic progress and obtain straightforward feedback regarding therapy skills. Consequently, the person-based

FIGURE 2.6 Psychotherapy can take a long time.

therapy process can be very time-consuming (see Figure 2.6), requiring numerous one-on-one sessions between professional therapist and client.

In contrast, behavior-based psychotherapy was designed to be administered by individuals with minimal professional training. From the start, the premise was to reach people where behavior-related problems occur—in the home, school, rehabilitation institute, and workplace—and teach parents, teachers, supervisors, friends, or coworkers the behavior-based techniques most likely to work under certain circumstances.

More than five decades of research have shown convincingly that this on-site approach is cost effective, primarily because behavior-change techniques are straightforward and relatively easy to administer, and because intervention progress can be readily monitored by the ongoing observation of target behaviors. By obtaining objective feedback on the impact of a particular intervention technique, a behavior-based process can be continually refined.

2.8.1 Integrating Approaches

A common perspective, even among psychologists, is that humanists and behaviorists are complete opposites. Behaviorists are considered cold, objective, and mechanistic, operating with minimal concern for people's feelings. In contrast, humanists are thought of as warm, subjective, and caring, with limited concern for directly changing another person's behavior or attitude.

Given the foundations of humanism and behaviorism, it is easy to build barriers between person-based and behavior-based perspectives, and to assume you must follow one or the other when designing an intervention process. In fact, most consultants who address the human dynamics of OSH market themselves as using one or the other approach, but not both. However, several scholars have promoted a combination of principles from humanism and behaviorism—humanistic behaviorism—for more effective and synergistic intervention impact (e.g., Dinwiddie, 1975; Thoresen, 1972; Kanfer & Phillips, 1970).

Relatedly, few, if any, students in introductory psychology courses read or hear the term "humanistic behaviorism." Instead, most introductory psychology textbooks emphasize distinct differences between humanistic and behavioral approaches to clinical therapy (Geller, 2020).

This text illustrates beneficial ways to integrate principles from humanism and behaviorism in order to optimize the beneficial impact of an intervention designed to improve one or more human dynamics of OSH. It is my firm belief that these approaches need to be integrated in order to truly understand the psychological science of safety and to achieve and sustain a TSC.

2.9 IN SUMMARY

Theory and basic principles are needed to organize research findings and guide our approaches to improve the OSH of an organization. Similarly, the vision for a TSC incorporated into a mission statement is invaluable for guiding the development of action plans to achieve an injury-free organization.

When employees understand and accept the mission statement and guiding principles, they become more involved in the mission. The action plan will not be viewed as one more flavor of the month but as interdependent responsibilities reflecting those principles that are relevant for achieving shared goals. Indeed, the workforce will help design, implement, and evaluate the action plans. This collaborative process is crucial for cultivating a TSC.

A basic principle introduced in this chapter is that the safety performance of an organization results from the dynamic interaction of environment, behavior, and person factors. The behavior and person dimensions represent the human dynamics of OSH and reflect two divergent approaches to understanding the psychological science of injury prevention.

Figure 2.7 summarizes the distinction between person-based and behavior-based psychology and shows that both approaches contribute to understanding and helping people. Both the internal and external dimensions of people are covered in this book as they relate to improving OSH.

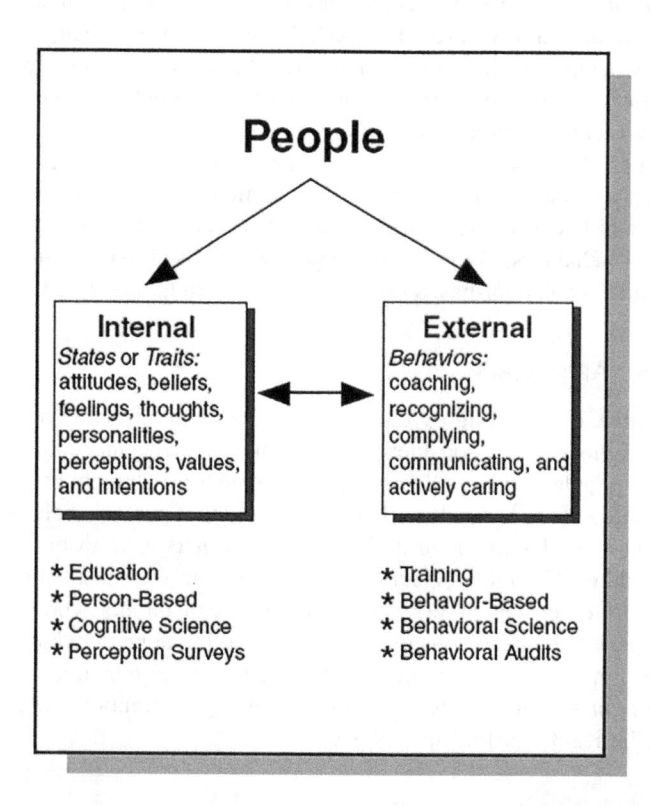

FIGURE 2.7 The internal and external dynamics of people determine the success of a safety process.

Profound knowledge on the person side comes from cognitive science, whereas the behavior-based approach is founded on behavioral science.

This text provides education by enhancing your knowledge and inspiring useful thinking about the human dynamics of OSH. Effective training can be accomplished by applying the observation and feedback techniques detailed later in Part 4 to improve your own or someone else's behavior.

Taken alone, the behavior-based approach is more cost effective than the person-based approach in cultivating large-scale improvement. However, behavior-based OSH can only be optimized if the workforce believes in the behavior-based principles and willingly applies them to achieve the mutual interdependent OSH mission. This requires a person-based approach. Therefore, to achieve a TSC, you need to integrate person-based and behavior-based psychology—humanistic behaviorism. This book shows you how to meet that challenge.

3 Paradigm Shifts for a Total Safety Culture

This chapter outlines ten perspectives you need to adopt in order to exceed current levels of OSH excellence and reach the ultimate goal—a Total Safety Culture. The traditional three Es of safety management—engineering, education, and enforcement—have only gotten us so far. A TSC requires understanding and applying three additional Es—empowerment, ergonomics, and evaluation.

> Mindsets are yesterday—mind growth is tomorrow.
>
> (Joe Batten)

Occupational safety and health (OSH) in industry has improved dramatically over the last century. Let's examine the evolution of injury prevention to understand how this was accomplished. The first systematic OSH-related research began in the early 1900s and focused on finding the psychological causes of workplace injuries. It assumed people were responsible for most mishaps and injuries, usually through mental errors caused by anxiety, attitude, fear, stress, personality, or another emotional state. The prevention of injuries was typically attempted by "readjusting" one's attitude or disposition, usually through supervisor counseling or discipline.

This so-called "psychological approach" held that certain individuals were "accident prone." By removing those workers from risky jobs or by disciplining them to correct their attitude or disposition, it was believed that injuries could be prevented. As presented in Chapter 1, this focus on "accident proneness" has not been effective, partly because reliable and valid measurement procedures are not available. Also, the person factors contributing to accident proneness are probably not consistent characteristics or traits within people but vary from time to time and situation to situation.

3.1 THE OLD THREE ES

Enthusiasm for the early "psychological approach" waned because of the difficulty in measuring its impact. In addition, the seminal research and scholarship of William Haddon (1968) suggested that engineering changes held the most promise for large-scale and long-term reductions in injury severity.

As the first Administrator of the National Highway Safety Bureau—now the National Highway Traffic Safety Administration (NHTSA)—Dr. Haddon was able to turn his theory and research into the first federal automobile safety standards. Haddon believed injury is caused by delivering excess energy to the body and that injury prevention depends on controlling that energy. The injury-prevention focus shifted to engineering and epidemiology and resulted in developing personal protective equipment (PPE) for work and recreational environments, as well as standards and policy regarding the use of PPE. Haddon's basic theory eventually led to collapsible steering wheels, padded dashboards, head restraints, and air bags in automobiles.

This brief history of the safety movement in the U.S. explains why engineering has been the dominant paradigm in OSH, with secondary emphasis on two additional "Es"—education and enforcement. For several decades, the basic protocol for reducing injury was to:

1. Engineer the safest equipment, environmental settings, and protective devices.
2. Educate people regarding the use of the engineering interventions.
3. Enforce compliance with the recommended safe work practices.

Those three Es have dramatically reduced injury severity in the workplace, at home, and on the road. Consider, for example, motor vehicle safety. The Government Accounting Office has estimated conservatively that the early automobile safety standards ushered through Congress by William Haddon had saved at least 28,000 American lives by 1974. In addition, the state laws passed in the 1980s requiring the use of vehicle safety belts and child safety seats have saved countless more lives. Many more lives would be saved and injuries avoided from vehicle crashes if more people buckled up and used child safety seats appropriately.

The current rate of safety-belt use in the U.S. is about 80 percent, a dramatic improvement from the 15 percent prior to statewide interventions, including belt-use laws, campaigns to educate people about the value of safety-belt use, and large-scale enforcement blitzes by local and state police officers.

Now, let's turn our attention to industry. I have worked with many corporate safety professionals over the years who say their plant's safety performance has reached a plateau. Yes, their overall safety record is vastly better than it once was, but continuous improvement is elusive. Many attempts to take OSH to the next level have not paid off. The old "three Es" paradigm will not get you there. A certain percentage of people keep falling through the cracks. Keep on doing what you're doing, and you will keep on getting what you're getting. As I heard W. Edwards Deming (1991) say many times, "Goals without method, what could be worse?"

3.1.1 THREE OTHER ES

This text discusses three additional Es—ergonomics, empowerment, and evaluation. I certainly do not suggest abandoning tradition. We need to maintain a focus on engineering, education, and enforcement strategies. However, to get beyond current plateaus and reach new heights in OSH excellence, we need to attend more competently to the psychological science of injury prevention. These three additional Es suggest specific directions for improving OSH.

3.1.1.1 Ergonomics

As discussed in Chapter 1, ergonomics requires careful study of relations between environment and behavior, as well as developing action plans (such as equipment work orders, safer operating procedures, and training exercises) to avoid possible acute or chronic injury from environment-behavior interactions. This requires consistent and voluntary participation by employees who perform those behaviors in the various work environments. These are usually line operators or hourly workers in an organization, and their participation will happen when those individuals feel empowered. Throughout this book, strategies for developing an empowered work culture are exemplified, as well as procedures for involving employees in ergonomic interventions.

3.1.1.2 Empowerment

Some operational definitions of the traditional three Es for safety (especially enforcement) have been detrimental to employee empowerment. Many supervisors have translated "enforcement" into a strict "punitive" approach, and the result has turned off many employees to safety programs. Those workers may do what is required, but no more. Some individuals who feel especially controlled by top-down safety regulations might try to beat the system, and success will likely bring a sense of gratification or freedom, as illustrated in Figure 3.1.

This principle is discussed in more detail in Chapter 15, especially as it relates to developing and implementing effective behavior-focused interventions. At this point, you need to understand

FIGURE 3.1 Some top-down rules have undesirable side-effects.

that some types of enforcement can inhibit empowerment and should be reconsidered and refined. Some paradigms must change—the theme of this chapter.

3.1.1.3 Evaluation

The third new E-word essential to achieving a TSC is evaluation. Without appropriate feedback or evaluation, practice does not make perfect. Thus, we need the right kind of evaluation processes. Later, especially in Chapter 15, practical procedures are provided for conducting an effective and comprehensive evaluation process. For now, it is important to understand that some traditional methods of evaluation actually decrease or stifle empowerment. This calls for changing some safety-measurement paradigms. In fact, basic theory from person-based and behavior-based psychology suggests ten safety paradigm shifts, which provide a new set of guiding principles for achieving and sustaining a TSC.

3.2 SHIFTING PARADIGMS

Ten basic changes in OSH beliefs, attitudes, and perceptions are needed to cultivate and achieve a TSC. These paradigm shifts implicate new principles and intervention procedures that require different behaviors and attitudes among managers and wage workers. If these paradigm shifts occur, empowerment will increase throughout the work culture, enabling the achievement of a TSC.

3.2.1 From Government Regulation to Corporate Responsibility

Many safety activities and programs in U.S. industry are driven by the Occupational Safety and Health Administration (OSHA) or the Mine Safety and Health Administration (MSHA) rather than by the employers and employees who directly benefit from a successful OSH process. In other words, many industry personnel do "safety stuff" because the government requires it—not because it was their idea and initiative.

Paradigm Shifts for a Total Safety Culture

FIGURE 3.2 Top-down control can stifle creativity.

People are more motivated and willing to go beyond the call of duty when they are achieving their own self-initiated goals. Ownership, commitment, and proactive behaviors are less likely when individuals are working to avoid missing goals or deadlines set by someone else. This statement is intuitive and reflected in Figure 3.2. Just compare your own motivation when working for personal gain versus someone else's gain or when working to earn a reward versus to avoid a penalty.

The language used to define safety programs and activities influences personal participation. Remember, we can act ourselves into an attitude. So, it makes sense to talk about safety as a company mission that is owned and achieved by the very people it benefits. A safety process is not intended to benefit federal regulators. Let's work to achieve a TSC for the right reasons.

3.2.2 From Failure Focused to Achievement Focused

If you strive to meet someone else's goals rather than your own, you will probably develop an attitude of "working to avoid failure" rather than "working to achieve success." We are happier and more self-motivated when achieving success than when avoiding failure. If you have a choice between earning positive consequences (rewards) or avoiding negative consequences (penalties), you will probably choose the positive-consequence situation. Moreover, when you feel controlled by negative consequences, you often procrastinate and take a reactive rather than a proactive stance.

Figure 3.3 illustrates what I mean. The runner will surely start running, but how long will he run? Will he keep running as fast when the coach is not around to threaten a negative consequence for not running? Will the runner practice on his own to improve his running skills? Will he hold himself accountable to be the best he can be on the running track? A "yes" answer to these questions will only occur if the runner can put himself in an achievement-oriented mindset. This is difficult in the enforcement context established by the coach.

This principle helps explain why employees typically give more continuous and proactive attention to productivity than to OSH. Productivity goals are typically stated in achievement

FIGURE 3.3 Working to avoid failure stifles self-motivation.

terms, and gains are tracked and recorded as individual or team accomplishments, sometimes followed by rewards or positive recognition.

In contrast, safety goals are most often stated in failure-avoidance terms. How many times have you heard, "We will reach our safety goal after another month without a lost-time injury"? Does "keeping score" for OSH imply tracking and recording losses or injuries?

Measuring safety with only records of injuries not only limits evaluation to a reactive stance, but it also sets up a negative motivational system that is apt to take a back seat to the positive achievement system used for productivity. Giving safety an achievement perspective (like production quality and quantity) requires a different scoring system, as indicated by the next paradigm shift.

3.2.3 FROM OUTCOME FOCUSED TO BEHAVIOR FOCUSED

Companies are frequently ranked according to their OSHA recordables and lost-time injuries. Within companies, work groups or individual workers earn safety awards according to outcomes—those with the lowest numbers win. Offering incentives for fewer injuries, for instance, can often reduce the *reported* numbers of injuries while not improving safety. Pressure to reduce outcomes without changing the process (or the ongoing safety-related behaviors) often causes employees to cover up their injuries.

Have you heard of an injured employee being driven to work each day to sign in and then promptly returned to the hospital or home to recuperate? This keeps the outcome numbers low but does more harm than good to the corporate culture. Likewise, failure to report even a minor first-aid case prohibits key personnel from correcting the factors that led to the incident.

A misguided emphasis on outcomes rather than process is illustrated in Figure 3.4. Although the idea of a dead person receiving a safety reward is clearly ridiculous, that type of incentive/reward process has been quite common in American industry. Such programs often bring down numbers by influencing the reporting of injuries, but rarely do they benefit the safety processes that determine the results or the outcome.

Paradigm Shifts for a Total Safety Culture

> JOE GETS THE SAFETY PRIZE AGAIN. HE WENT ANOTHER 30 DAYS WITHOUT AN OSHA RECORDABLE.

FIGURE 3.4 Safety reward programs should pass the "dead-man's test."

A scoring system based on what people do for safety is achievement focused and thereby encourages action for injury prevention. This positions safety in the same motivational framework as productivity.

Chapter 11 explains principles for establishing an incentive/reward process to motivate the kinds of safety processes that enhance desirable outcomes. For now, just recognize and appreciate the advantage of focusing on achieving process improvements over working to avoid failure. This is especially true if a failure-oriented goal is remote, such as a plant-wide reduction in injuries, which might be perceived as uncontrollable.

Safety can be comparable to productivity quality and quantity if it is recorded and tracked with an achievement score that is perceived by employees as directly controllable and attainable. This occurs with a focus on the safety processes that can decrease an organization's injury rate, as well as an ongoing measurement system that regularly tracks safety accomplishments and displays them to the workforce. This text shows you how to make that happen.

3.2.4 FROM TOP-DOWN CONTROL TO BOTTOM-UP INVOLVEMENT

As discussed when introducing three additional E-words, a TSC requires continual involvement from operations personnel, such as hourly workers. After all, those individuals are the employees who know the locations of various safety hazards, and observe when particular at-risk behaviors occur. Also, those individuals can have the most influence in supporting safe behaviors and correcting at-risk behaviors and conditions. In fact, the ongoing processes involved in cultivating a TSC need to be supported from the top but driven from the bottom. This is more than employee participation; it is employee ownership, commitment, and empowerment.

3.2.5 FROM RUGGED INDIVIDUALISM TO INTERDEPENDENT TEAMWORK

An employee-driven OSH process requires teamwork founded on interpersonal trust, synergy, and win/win contingencies. However, from childhood most of us have been taught an individualistic, win/lose perspective, supported by such popular slogans as "You have to blow your own

FIGURE 3.5 The U.S. culture promotes more independence than interdependence.

horn," "Nice guys finish last," "No one can fill your shoes like you," and "The squeaky wheel gets the oil." Furthermore, as shown in Figure 3.5, grades in school, the legal system, and many athletic competitions orient us to think win/lose independence rather than win/win interdependence. That is why a true team approach to OSH does not come easily.

3.2.6 From a Piecemeal to a Systems Approach

The achievement of a TSC requires a systems approach, including balanced attention to all aspects of the corporate culture. Deming (1986b, 1993) emphasized that total quality can only be achieved through a systems approach, and of course the same is true for OSH. As discussed earlier in Chapter 2, three basic domains need attention when designing and evaluating OSH processes, and when analyzing the potential contributors to close calls and unintentional injuries.

1. Environment factors such as equipment, tools, machines, housekeeping, heat/cold climate, and engineering.
2. Person factors such as employees' knowledge, skills, abilities, intelligence, motives, and personality.
3. Behavior factors such as complying, coaching, recognizing, communicating, and "actively caring."

Two of those system variables are human factors. Such human dynamics generally receive less attention than the environment, mostly because it is more difficult to visibly measure the outcomes of efforts to change human factors. Some human-factor processes focus on behaviors (as in behavior-based safety); other psychological science processes address attitudes (as in a person-based approach). A TSC integrates these two approaches for OSH achievement.

3.2.7 From Fault Finding to Fact Finding

Blaming an individual or group of individuals for an injury-producing incident is not consistent with a systems approach to OSH. Instead, an injury or close call provides an opportunity to

Paradigm Shifts for a Total Safety Culture

gather facts from all aspects of the system that could have contributed to the incident. However, most evaluations of close calls or injuries are incomplete, and are much less informative than they could be. Part of the problem here is the very term we use to describe the process—*accident investigation*.

"Accident investigation" is a common phrase in OSH, but what does it mean? The term "accident" implies "a chance occurrence" outside your immediate control. When a young child has an "accident" in his/her pants, we presume s/he was not in control. S/he could not help it.

And what about the word "investigation"? That term seems to imply a hunt for some single cause or person to blame for a particular incident, as in "criminal investigation." How can we promote fact-finding over fault-finding with a term like "investigation" defining our job assignment?

To learn how to prevent injuries from an analysis of an incident, we need to approach the task with a different mindset. It is not an "accident investigation;" it is a "close call or injury analysis." This change in language can get more employee participation in the process and reap more benefits as a consequence.

Moreover, do you really believe there is a single "root cause" of an unintentional incident, whether a close call, property damage, or a personal injury? If you consider the interactive effects of environmental, behavioral, and person factors that influence safety-related performance, you must answer "No" to that rhetorical question. Environmental factors include tools, equipment, engineering design, management systems, housekeeping, and climate. Then we have the actions or behaviors of everyone related to the mishap, as well as the attitudes, perceptions, dispositions, and personality characteristics of those individuals.

Given the dynamic interdependency of those OSH-related factors, searching for a single root cause is irrational. Furthermore, cause-and-effect solutions cannot be found from information obtained from surveys, group discussions, or personal interviews. Only correlational data can be attained from such assessments, not causal contingencies. More importantly, analyzing the potential contributing factors to an unintentional mishap instead of conducting a fault-finding "accident investigation" is needed to explore strategies for preventing injuries and cultivating the achievement of a TSC.

FIGURE 3.6 Technology cannot always substitute for personnel.

3.2.8 From Reactive to Proactive

Analyzing events preceding an incident, whether a close call or an injury, demonstrates the need to think and act proactively. Unfortunately, a proactive stance is extremely difficult to maintain, especially in a corporate culture that is increasingly complex and demanding. There is seemingly a higher and higher price tag on "free time." With barely enough time to react sufficiently to daily crises, how can we find time to be proactive?

Proactivity is especially challenging within the context of downsizing, disguised as "re-engineering" in many work cultures. The worker in Figure 3.6 is barely able to react effectively to daily crises. How can he be expected to think ahead and be proactive? There are no quick-fix answers to this challenging question, but injury prevention requires us to find solutions. This text provides theory, procedures, and tools to guide long-term continuous improvement. Thus, we need to accept the next paradigm shift.

3.2.9 From Quick Fix to Continuous Improvement

"Proactive" can be substituted for "reactive" only with a systems perspective and an optimistic attitude of continuous improvement through increased and sustained employee involvement. Understanding the human dynamics of OSH can be an invaluable aid here. The principles and procedures described in this book will enable you to influence incremental changes in cognitions, attitudes, and behaviors relevant for preventing an injury to yourself and to others. That reflects a proactive, continuous-improvement paradigm, which will surely improve your OSH performance.

3.2.10 From Priority to Value

"Safety is our priority." This is probably the most common safety slogan found in workplaces and voiced by safety mangers. I have seen signs, pens, buttons, hats, T-shirts, and notepads with this message. No wonder safety and health professionals are surprised when I proclaim that safety should *not* be a priority. To justify my case, I offer the following explanation.

Think about a typical workday morning. We usually follow a prioritized agenda, often a standard routine, before traveling to work. Some people eat a healthy breakfast, read the morning newspaper, take a shower, and wash dishes. Others wake up early enough to go for a morning jog before work. Some grab a roll and a cup of coffee and leave their home in disarray until they get back in the evening.

In each of those scenarios, the agenda—the priorities—are different. Yet, there is one common activity. It is not a priority but a basic value. Do you know what it is?

One morning you wake up late. Perhaps your alarm clock failed. You have only 15 minutes to prepare for work. Your morning routine changes drastically. Priorities must be rearranged. You might skip breakfast, a shower, or a shave. Yet every morning schedule still has one item in common. It is not a priority, capable of being dropped from a routine due to time constraints or a new agenda. No, this particular morning activity represents a value which we have been taught as infants, and it is never compromised. Have you guessed it by now? Yes, this common behavior included in everyone's morning routine, regardless of time constraints, is "getting dressed."

That simple scenario shows how circumstances can alter behavior and priorities. Actually, labeling a behavior a "priority" implies that its order in a hierarchy of daily activities can be rearranged. How often does this happen at work? Does safety sometimes take a "back seat" when the emphasis is on other priorities such as production quantity or quality?

3.2.11 ENDURING VALUES

It is human nature to shift priorities, or behavioral hierarchies, according to situational demands or contingencies. But values remain constant. The early morning anecdote illustrates that the activity of "getting dressed" is a value that is never dropped from the routine. Shouldn't "working safely" hold the same status as "getting dressed"? Safe work practices should occur regardless of the demands of a particular day.

Safety should be a value linked to every activity or priority in a work routine. Safe work should be the enduring norm, whether the current focus is on quantity, quality, or cost effectiveness as the "number one priority."

The ultimate aim of a TSC is to make safety an integral component of all performance, regardless of the task. Safety should be more than the behavior of "using PPE," more than "locking out power" and "checking equipment for potential hazards" and more than "practicing good housekeeping." Safety should be an unwritten rule—a social norm—that workers follow regardless of the situation. It should become a value that is never compromised.

3.3 IN SUMMARY

This chapter described ten shifts in perspective or mindset needed to go beyond current levels of OSH excellence. The first nine paradigm shifts could be considered goals for achieving a TSC. The tenth paradigm shift—making safety a value—is not something that can be measured and tracked. It is the ideal vision for our actively-caring-for-safety mission.

Let's consider how these new paradigms fit together. Your OSH achievement process should be a company responsibility, not a regulatory obligation. It should be achievement oriented with a focus on OSH-related behavior, supported by all managers and supervisors but driven by the line workers or operators through interdependent teamwork. A systems approach is needed, which leads to a fact-finding perspective, a proactive stance, and a commitment to continuous improvement.

These new perspectives reflect new principles to follow, and new procedures to develop and implement. This "new OSH mindset" will lead to different perceptions, attitudes, behaviors, and even values. Ultimately, the tenth paradigm shift can be reached. When safety goes from priority to value, it will not be compromised at work, at home, or on the road. Numerous injuries will be prevented and lives saved every day. This vision should motivate each of us to actively care for safety at home, at work, and everywhere in between. This book was designed to help you define your role in achieving and sustaining a TSC at work and at home.

Part Two

Human Barriers to Safety

4 The Complexity of People

Safety is usually a continuous fight with human nature. This chapter explains why. Understanding this basic point will lead to less victim blaming and fault finding when analyzing an injury. Instead, you will be able to find factors in the system that can be changed in advance in order to prevent injuries at work, at home, and throughout the community.

What lies behind us and what lies before us are small matters compared to what lies within us.

(Ralph Waldo Emerson)

"All injuries are preventable."
"It is human nature to work safely."
"Safety is just commonsense."
"Safety is a condition of employment."

Read those familiar statements and you get the idea that working safely is easy or natural. Nothing could be further from the truth.

In fact, it is often more convenient, more comfortable, more expedient, and more common to take risks than to work safely. Plus, past experience usually supports our decisions to choose the at-risk behavior, whether we are working, traveling, or playing. Thus, we are often engaged in a continuous fight with human nature to motivate ourselves and others to avoid those risky behaviors and maintain safe behaviors.

Let's consider what holds us back from choosing the safe way, whether following safe operating procedures, driving our automobile, or using personal protective equipment.

4.1 FIGHTING HUMAN NATURE

When I have asked safety professionals, corporate executives, or hourly workers what causes work-related injuries, I got long and varied lists of factors. However, each list is quite similar. After all, everyone experiences events, attitudes, demands, distractions, responsibilities, and circumstances that get in the way of performing a task safely.

Most of us have been in situations where we were not sure how to perform safely. Perhaps we lacked training. Maybe the surrounding environment was not as safe as it could be. Demands from a supervisor, coworker, or friend put pressure on us to take a risky shortcut. Perhaps, it was inconvenient or uncomfortable to follow all of the safety procedures.

It is possible our physical condition—fatigue, boredom, drug impairment—influenced at-risk behavior. There are other factors. Have you ever been unsafely distracted by external stimuli, like another person's presence or by an internal person-state, like personal thoughts or emotions? Can you remember a time when you just did not feel like taking the extra time to be safe?

I am sure you have experienced the "macho" attitude from yourself or from others that, "It will never happen to me." Fortunately, it is rare that an injury follows unsafe behavior. The attitude, "It won't happen to me," is usually supported or reinforced by our actual experiences. Risk taking is rarely punished with an injury or even a close call. Instead, at-risk behavior is consistently rewarded with convenience, comfort, and time saved.

This creates a vicious cycle. The rewards of risky behavior can influence us to take more chances. As you gain experience at work you often master dangerous shortcuts. Because those at-risk behaviors are not followed by a close call or an injury, they remain unpunished, and they persist.

This basic principle of human nature is naturally or intrinsically reinforced throughout our lives, and runs counter to the safety-related efforts of individuals, groups, and organizations. It explains why promoting OSH is the most difficult ongoing challenge in the workplace. The reality is that injuries really do happen to the "other guy."

4.1.1 Dimensions of Human Nature

The factors contributing to a work injury can be categorized into three areas.

1. Environment factors.
2. Person factors.
3. Behavior factors.

That is the "Safety Triad" introduced in Chapter 2. The most common reaction to an injury is to correct something about the environment—modify or fix equipment, tools, housekeeping, or an environmental hazard.

Often the incident/injury report includes some mention of person factors, like the employee's knowledge, skills, ability, intelligence, motives, or personality. These factors are typically translated into general recommendations.

"The employee will be disciplined."
"The employee will be retrained."

That kind of vague attention to the critical human dynamics of a work injury shows how frustrating and difficult it can be to deal with the psychology of OSH—the person and behavior sides of the Safety Triad. The human factors contributing to an injury are indeed complex, often unpredictable and uncontrollable. This justifies a conclusion that all injuries cannot be prevented.

The acronym BASIC ID reflects the complexity and uncontrollability of human nature. As depicted in Figure 4.1, each letter represents one of seven human dimensions of an individual. Here is a simple scenario that highlights the potential impact of various human factors on an injury, and the challenging need to address the human dynamics of OSH.

> *Dave, an experienced and skilled craftsman, works rapidly to make an equipment adjustment while the machinery continues to operate. As he works, production-line employees watch and wait to resume their work. Dave realizes all too well that the sooner he finishes his task, the sooner his coworkers can resume quality production. So, he does not shut down and lock out the equipment power. After all, he has adjusted this equipment numerous times before without locking out, and he has never gotten injured.*
>
> *A morning argument with his teenage daughter pervades Dave's thoughts as he works, and suddenly he experiences a close call. His late timing nearly results in his hand being crushed in a pinch point.*
>
> *Removing his hand just in time, Dave feels weak in his knees and begins to perspire. This stress reaction is accompanied by a vivid image of a crushed right hand. After gathering his composure, Dave walks to the switch panel, shuts down and locks out the power. Then he lights up a cigarette. He thinks about this scary event for the rest of the day, and talks about that close call with fellow workers during his breaks.*

That brief episode illustrates each of the psychological dimensions represented by BASIC ID (see Figure 4.1), and demonstrates the complexity of human performance.

Behavior is illustrated by observable actions such as adjusting equipment, pulling a hand away from the moving machinery, lighting up a cigarette, and talking to coworkers.

The Complexity of People

```
B ehavior
A ttitude
S ensation
I magery
C ognition

I nterpersonal
D rugs
```

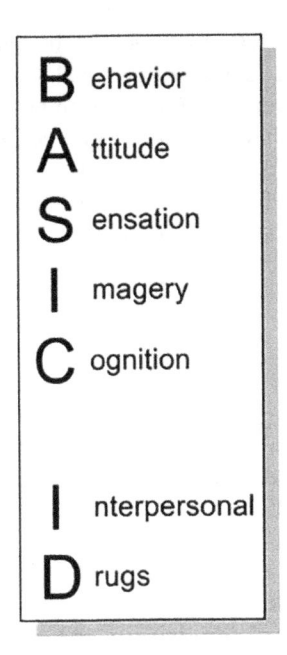

FIGURE 4.1 The acronym BASIC ID reflects the complexity of people and potential contributions to an injury.

Dave's **attitude** about work was fairly neutral at the start of the day, but immediately following his close call he felt a rush of emotion. His attitude toward "energy control and power lockout" changed dramatically, and his commitment to the locking-out protocol increased after relating his close call to friends.

Sensation is evidenced by Dave's dependence on visual acuity, hand–eye coordination, and a keen sense of timing when adjusting the machinery. His ability to react quickly to the dangerous situation prevented severe pain and potential loss of valuable touch sensation.

Imagery occurred after the close call when Dave visualized a crushed hand in his "mind's eye," and this contributed to the significance and distress of the incident. Dave will probably experience that mental image periodically for some time to come. This will motivate him to perform appropriate lockout procedures, at least for the immediate future.

Cognition or "self-talk" about the morning argument with his daughter may have contributed to the timing error that resulted in the close call. Dave will probably remind himself of this incident in the future, and those cognitions may help activate proper lockout behavior.

Interpersonal refers to the other people in Dave's life who contributed to his close call and will be influential in determining whether he initiates and maintains appropriate lockout practices. For example, it was the interpersonal discussion with his daughter that occupied his thoughts or cognitions before the close call. The presence of production-line workers influenced Dave through subtle peer pressure to quickly adjust the equipment without following the lockout protocol. Those onlookers may have distracted Dave from the task, or they could have motivated him to show off his adjustment skills. After Dave's close call, his interpersonal discussions were therapeutic, helping him relieve his distress and increase his personal commitment to OSH.

Drugs in the form of caffeine from morning coffee may have contributed to Dave's timing error. The extra cigarettes Dave smoked as a "natural" reaction to distress also had physiological consequences, which could have been reflected in Dave's subsequent behavior, attitude, sensation, or cognition.

Dave's lesson shows how human nature interacts with environmental factors to influence at-risk work practices, close calls, and sometimes personal injuries. It is relatively easy to control the environmental factors. As explained in Section 3 on behavior-based safety (BBS), it is feasible to measure and control the behavior factors. However, the complex person factors, represented by the BASIC ID acronym, are quite elusive. Psychological science provides insights here, and this information can benefit OSH. Let's discuss further various aspects of human nature that can make safety achievement especially challenging.

4.2 COGNITIVE FAILURES

"All injuries are preventable." I have heard this said so many times that it seems to be a slogan or personal affirmation that safety pros use to keep themselves motivated. I suspect some readers will resist any challenge to that ideal. Their optimism is certainly appreciated, and there is no harm if such perfectionism is kept to oneself. However, sharing that belief with others can actually hinder the achievement of a TSC.

If a common workplace slogan declares all injuries preventable, workers may be reluctant to admit they were injured or had a close call. After all, if all injuries are preventable and I have an injury, I must be a real "jerk" for getting hurt.

Combine that slogan with a goal of zero injuries and a reward for not having an injury and human nature will dictate covering up an injury if possible. Also, as explained in the next chapter, claiming that all injuries are preventable can reduce the perceived risk of the situation. This can create the notion that "it will not happen to me"—a perception that can increase the probability of at-risk behavior and an eventual injury.

Eliminate the "all injuries are preventable" slogan from your OSH discussions. The most important reason to drop that assertion is that most people do not believe it anyway. They have been in situations where all the factors contributing to the close call or injury could not have been anticipated, controlled, or prevented. The most uncontrollable factors are the person-based or internal subjective dimensions of people. Consider, for example, the role of "cognitive failures."

Donald Norman (1988) classified various types of cognitive failure according to a particular stage of routine thinking and decision-making. More specifically, consider that we continually take in, process, and react to information in our surroundings. Almost everything we do results from this basic information-processing cycle.

We sense a stimulus (input), we evaluate the stimulus and plan a course of action (interpretation and decision-making), and then we execute a response (output). Unintentional cognitive errors usually occur at the input and output stage of information processing. Judgment errors and calculated risks occur at the middle cognitive stage—interpretation and decision-making.

4.2.1 Capture Errors

Have you ever started traveling in one direction (like to the store) but suddenly find yourself on a more familiar route (like on the way to work)? How many times have you borrowed someone's pen to write a note or sign a form, and later found that pen in your pocket? Norman calls those "capture errors," because a familiar activity or routine seemingly "captures" you and takes over an unfamiliar activity. This error seems to occur at the execution stage of information processing, but it also involves misperception or inattention to relevant stimuli, as well as the absence of conscious judgment or decision-making.

How does this error slip into the work routine? Have you ever started a new task and found yourself using old habits? Has a change in PPE requirements influenced this kind of human

The Complexity of People

error? It seems reasonable that a routine way of doing something (even at home) could "capture" your execution of a new work process and lead to this type of cognitive failure and an injury.

4.2.2 Description Errors

These "brain cramps" occur when the descriptions or locators of the correct (safe) and incorrect (at-risk) executions are similar. For example, I have thrown a tissue in the clothes hamper instead of the waste can, even though the clothes hamper is not next to a trash receptacle. On a few occasions, I have actually thrown dirty clothes in the trash can, and once I threw a sweaty T-shirt in the toilet. According to Norman, the similar characteristic of these three items—a large oval opening—led to those errors.

Do you have any switches in your work setting that are similar and nearby but control different functions? How unsafe would it be to throw the wrong switch? Many control panels are designed with consideration of the possibility of this error. Switches or knobs controlling incompatible functions are not located in close proximity with one another, and they often look and feel distinctly different for quick visual and tactile discrimination.

Thus, it might be useful to evaluate your work setting with regard to the need for different behaviors with similar descriptions or locations.

4.2.3 Loss-of-Activation Errors

Have you ever walked into a room to do something or to get a certain object, but when you got there, you forgot what you were there for? You think hard but just cannot remember. Then, you go back to the first room and suddenly you remember what you wanted to do or get in the other room. What happened here?

This cognitive failure is commonly referred to as "forgetting." Norman refers to it as "loss-of-activation," because the cue or the activator that got the behavior started was lost or forgotten. This happens whenever you start an activity with a clear and specific goal, but after you get engaged in the task, you lose sight of the goal. You might, in fact, continue the task but with little awareness of the rationale for continued progress toward achieving a goal.

When people tell you they already know what to do with statements like "Stop harping on the same old thing," you can say you are just actively caring by trying to prevent a "loss-of-activation" error. You will never know how many of those cognitive failures you will prevent, but you can motivate yourself to keep activating by reflecting on your own experiences with this type of "brain cramp." Then, consider the large number of people in your work setting who make similar unintentional errors every day.

4.2.4 Mode Errors

Mode errors are probable whenever we face a task involving multiple options or modes of operation. These errors are inevitable when equipment is designed to have more functions than the number of control switches available. In other words, when controls are designed for more than one mode of operation, you can expect occurrences of this error.

Over the years, I have owned a variety of digital watches with a stopwatch mode. Each watch has had a different arrangement of switches designed to provide more functions per control buttons. Therefore, the meaning of a button press depends upon the position of a mode switch. So, guess how many times I have pressed the wrong button and illuminated the dial, or reset the digital readout when I only wanted to stop the timer? Have you experienced the same kind of mode error, if not with a stopwatch, perhaps with the text editor of a personal computer? Airline pilots must be especially wary of this kind of error.

4.2.5 Mistakes and Calculated Risks

The four types of cognitive failures discussed so far—capture, description, loss-of-activation, and mode errors—are unintentional. Their sources are mostly at the input and output stages of information processing. The intermediate interpretation and decision-making stages are essentially uninvolved. Thus, the at-risk behavior resulting from those errors is unintentional. The person meant well. The plan was good, but the execution was unintentionally flawed.

Mistakes and calculated risks occur at the interpretation and decision-making stage of information processing. Here is where we interpret our sensory input and decide on a course of action. With mistakes, the individual was well intentioned regarding the ultimate outcome of getting the job done but used poor judgment in getting there.

While driving, have you ever turned right on to a main road into the path of an oncoming vehicle you had not seen, or whose speed you had misjudged? Have you ever miscalculated a parking space and scraped an adjoining vehicle? How many times have you planned a bad travel route and got caught in traffic congestion you could have avoided? Have you ever pressed the brake too quickly on a slippery road or pumped the brakes in an antilock system? Parking and braking are frequent and intentional driving behaviors, but under certain circumstances, they are mistakes.

Now suppose you do not buckle your safety belt. Perhaps you divide your attention between the road and some other task like map reading, text sending, cell-phone dialing, or Internet selection. You know that behavior is unsafe, but you decide to take a "calculated risk." Your judgment was faulty, as in a mistake, but unlike a mistake, your at-risk behavior was deliberate. Such behavior might seem rational because it is not followed by a negative consequence, and it is supported with perceived comfort, convenience, and/or efficiency.

4.3 INTERPERSONAL FACTORS

Our interpersonal relationships dramatically influence our thoughts, attitudes, and actions. How much of your time each day is dedicated to gaining the approval of others? Of course, we sometimes attempt to avoid the disapproval of others, be they a parent, spouse, work supervisor, or department head. As discussed in Chapter 3, we do not feel as good—or as "free"—when working to avoid failure or disapproval as when working to achieve success or approval. In both cases, other people are the cause of our motivation and behavior.

Consider for a moment the adversity many people go through to impress others. Have you ever followed orders you knew put yourself or others at some degree of risk? If something went wrong, it would not be your fault. It was not your responsibility; you were just following orders. Just like when we were kids and we got into trouble, we were quick to say, "It's not my fault, he told me to do it."

It is easy to see the relevance of these scenarios for OSH. People take risks on the job because others do the same, and sometimes workers blindly follow a supervisor's orders that could endanger them, other coworkers, or the environment. This reflects the interpersonal power of two principles of social influence—conformity and authority. Let's examine these interpersonal phenomena more closely to understand how they can be human barriers to safety. Then we can consider ways to turn these social influence factors around and use them to benefit OSH.

4.3.1 Peer Influence

The phenomenon of social conformity—depicted humorously in Figure 4.2—is certainly not new to any reader. We see examples of it every day, from the clothes people wear to how they communicate both verbally and in writing. We cannot overlook the power of conformity in influencing at-risk behavior. We have learned that peer pressure increases when more people are involved and when the group members are seen as relatively competent or experienced.

The Complexity of People 41

FIGURE 4.2 Peer pressure drives social conformity.

However, it is important to remember that one dissenter—a leader willing to ignore peer pressure and do the right thing—is often enough to prevent another person from succumbing to potentially dangerous conformity at work.

4.3.2 Power of Authority

Imagine you are among nearly 1000 participants in one of Stanley Milgram's 20 obedience studies at Yale University in the 1960s. You and another individual are led to a laboratory to participate in a human-learning experiment. First, you draw slips of paper out of a hat to determine randomly who will be the "teacher" and the "learner."

You get to be the teacher; the learner is taken to an adjacent room and strapped to a chair wired through the wall to an electric shock machine containing 30 switches with labels ranging from 15 volts—light shock—to 450 volts—severe shock. You sit behind this shock generator and are instructed to punish the learner for errors in the learning task by delivering brief electric shocks, starting with the 15-volt switch and moving up to the next higher voltage with each of the learner's errors. The scenario is depicted in Figure 4.3.

Complying with the researcher's instructions, you hear the learner moan as you flick the third, fourth, and fifth switches. When you flick the eighth switch (labeled 120 volts), the learner screams, "These shocks are painful," and when the tenth switch is activated, the learner shouts, "Get me out of here!"

At this point, you might think about stopping, but the researcher prompts you with words like, "Please continue—the experiment requires that you continue."

Increasing the shock intensity with each of the learner's errors, you reach the 330-volt level. Now you hear shrieks of pain—the learner pounds on the wall, then becomes silent. Still, the researcher urges you to flick the 450-volt switch when the learner fails to respond to the next question.

At what point will you refuse to obey the instructions? Milgram asked that question of a group of people, including 40 psychiatrists, before conducting his experiment. Most thought the sadistic

FIGURE 4.3 Participants experienced distress when giving electric shocks to peers.

game would stop soon after the learner indicated the shock was painful. Milgram and his associates (Milgram, 1963) were surprised that 65 percent of his actual subjects, ranging in age from 20 to 50, went along with the researcher's request right up to the last 450-volt switch.

Why did they keep following along? Did they figure out the learner was a confederate of the researcher and did not really receive the shocks? Did they realize they were being deceived in order to test their obedience?

No, the participants sweated, trembled, and bit their lips when giving the shocks, as shown in Figure 4.3. Some laughed nervously. Others openly questioned the instructions, but most did as they were told.

Milgram and associates learned more about the power of authority in further follow-up studies. Full obedience exceeded 65 percent, with as many as 93 percent flicking the highest shock switch, when:

- The authority figure—the researcher giving the orders—was in the room with the subject.
- The authority was supported by a prestigious institution, such as Yale University.
- The shocks were given by a group of "teachers" in disguise and remaining anonymous.
- There was no evidence of noncompliance—no other participant was observed disobeying the experimenter.
- The victim was depersonalized or distanced from the participant in another room.

From this research, Milgram (1974) concluded: "Ordinary people, simply doing their jobs and without any particular hostility on their part, can become agents in a terrible destructive process" (p. 605).

Let's apply that research to the workplace. As a result of social obedience or social conformity, people might perform risky acts or overlook obvious safety hazards, and put themselves and others at-risk. To say, "I was just following orders," reflects the obedience phenomenon, and "Everyone else does it!" implies social conformity or peer pressure.

To achieve a TSC, we need to realize the power of these two social influence principles—conformity and authority. Interventions capable of overcoming peer pressure and blind obedience are detailed in Part 4 of this book. Please note the crucial role of effective leadership. One person can make a difference—decreasing both destructive conformity and obedience—by deviating from the norm and setting a safe example. When a critical mass of individuals boards the "actively caring-for-safety bandwagon," you have achieved constructive conformity and obedience that support a TSC.

4.4 IN SUMMARY

We need to understand a problem as completely as possible and from many perspectives before we can solve it. In this chapter, we explored dimensions of OSH by considering the complexity of people. I attempted to convince you that human nature does not usually support OSH. The natural connections between behavior and its motivating consequences often result in some form of convenient, time-saving, and at-risk behavior. Consequently, to achieve a TSC, you should prepare for an ongoing fight with human nature.

Human barriers to OSH are represented by an acronym from clinical psychology—BASIC ID. The "C" (cognitions) and the second "I" (interpersonal) dimensions of this acronym, in particular, explain the special challenges of achieving a TSC. The phenomenon of cognitive failures reflects the shallowness—in fact, the potential danger—of the popular safety slogan, "All injuries are preventable." Conformity and obedience—two powerful principles from social psychological research—help us understand the individual, group, and system factors responsible for at-risk behavior and unintentional injury. Both of these social influence phenomena influence the kind of at-risk behavior depicted in Figure 4.4.

The human barriers to OSH discussed here can lead us to become more defensive and alert in hazardous environments. They also convey how difficult it is to find the factors contributing to a close call or an injury. Another psychological challenge to safety is explored in the next chapter with a discussion of the "S" (sensation) of BASIC ID.

FIGURE 4.4 Social conformity and obedience can inhibit safety-related behavior.

5 Sensation, Perception, and Perceived Risk

It is critically important to understand that perceptions of risk vary among individuals. We cannot dramatically improve OSH until people increase their perception of risk in various situations and reduce their overall tolerance for risk. In this chapter we explore the notion of selective sensation or perception, and then relate this concept to perceived risk and injury prevention. Several factors are discussed that influence whether employees react to workplace hazards with alarm, apathy, or something in between. Taken together, these factors shape personal perceptions of risk and illustrate why the job of improving OSH can be so challenging.

> What we see depends mainly on what we look for.
>
> **(John Lubbock)**

The "S" of the BASIC ID acronym introduced in Chapter 4 refers to sensation—a human dimension that influences our thinking, attitudes, emotions, and behavior. In grade school, we learned there are five basic senses we use daily to experience our world (we see, hear, smell, taste, and touch). Later we learned that our senses do not take in all of the information available in our immediate surroundings. Instead, we intentionally and unintentionally tune in and tune out certain features of our environment; thus, some potential experiences are never realized.

5.1 SELECTIVE SENSATION OR PERCEPTION

At the 1994 Professional Development Conference of the American Society of Safety Engineers (ASSE), the following instructions were printed on one-half of one page of the 40-page handout distributed to the audience of more than 350 individuals at the start of my two-hour presentation.

> You are going to look briefly at a picture and then answer some questions about it. The picture is a rough sketch of a poster of a couple at a costume ball. Do not dwell on the picture. Look at it only long enough to "take it all in" at once. After this, you will answer "yes" or "no" to a series of questions.

After the participants read the instructions, I presented the illustration depicted in Figure 5.1 for about five seconds. If you would like to experience the biased visual sensation (or perception) demonstrated to the ASSE audience, please read the instructions previously given and then look at Figure 5.1 for approximately five seconds. Then, answer the questions that I asked my ASSE audience.

1. Did you see a man in the picture?
2. Did you see a woman in the picture?
3. Did you see an animal in the picture? If so, what kind of animal did you see? What other details did you detect in the brief exposure to the drawing?
   ___ A woman's purse?
   ___ A man's cane?
   ___ A trainer's whip?

Sensation, Perception, and Perceived Risk 45

___ A fish?
___ A ball?
___ A curtain?
___ A test?

Practically everyone in the audience raised a hand to answer "yes" to the first question, and I suspect you also see a man in Figure 5.1. But only about one-half of the audience acknowledged seeing a woman in the drawing, and many said they had seen an animal. When I asked what type of animal, the common response heard across the room was "a seal." This drew many laughs, and the laughter got louder when I asked what else was quickly perceived in the illustration.

Several participants saw a woman's purse and a man's cane; others said they had seen a trainer's whip, a fish, and a beach ball. Some remembered seeing a curtain. Others saw part of a circus tent. What did you see in Figure 5.1?

"What's going on here?" I asked the ASSE audience. Why are we getting these diverse reactions to one simple picture? Some people speculated about environmental factors in the seminar room, including lighting, spatial orientation, and visual distance from the presentation screen. Others thought individual differences, including gender, age, occupation, and personal experiences "last night" could be responsible. Finally, someone asked whether the instructions printed in their handouts could have influenced the different perceptions. That was, in fact, the case.

Every handout included the same exact instructions except for a few words. One half of the handouts included all the words given previously; the other half of the handouts had the words "trained seal act" substituted for "couple at a costume ball." That was enough to make a marked difference in perceptions. Perhaps this makes perfect sense to you. Critical words in the instructions created expectations for a particular visual experience. I had set up my audience. Was your perception of Figure 5.1 influenced by the "set up" on the prior page?

FIGURE 5.1 Selective sensation can be demonstrated with this ambiguous drawing.

5.1.1 Biased by Context

Take a look at Figure 5.2. I suspect you have no difficulty reading the sentence as "The cat sat by the door," even though the symbol for "H" is exactly the same as the symbol for "A." It is a matter of context. The symbol was positioned in a way that influenced your labeling (or encoding) of the symbol. Likewise, the context or the environmental surroundings in our visual field influence how we see particular stimuli.

The same is true for our other senses—hearing, smelling, tasting, and touching. How we experience food, which involves the sensations of smell, touch, and taste, can be dramatically influenced by the atmosphere in which it is served. This is a basic rule of the restaurant business. Of course, other factors also bias food sensations, including hunger and past experiences with the same and similar food.

Now, take a look at Figure 5.3. What label do you give the man in the drawing on the left? The setting or context certainly influences your decision. The sign, keys, and uniform are cues that the man is probably a doorman. The environmental context in the drawing on the right leads to a different perception and label for the same person. Here, he is perceived as a policeman enforcing a safety policy.

Let's take our discussion of perception and apply it to the workplace. In this setting, perceptions of people can be shaped by equipment, housekeeping, job titles, and work attire. In fact, our own job title or work assignment can influence perceptions of ourselves, as well as affect our perceptions of others. This can dramatically influence how we interact with others, if we choose to interact at all.

It is important to recognize this contextual bias. Pick out someone with whom you communicate at work, and think how your relationship would be different in another setting. Would you

FIGURE 5.2 The context or circumstances surrounding a stimulus can influence how we perceive it.

FIGURE 5.3 The environmental context influences personal perception of the man in uniform.

Sensation, Perception, and Perceived Risk 47

still feel superior or inferior? The work setting has a way of turning individuals into numbers, depersonalizing them. This impression certainly can be misleading and might cause you to overlook someone's potential. In another setting, that same individual might feel empowered and be perceived as a competent leader.

5.1.2 BIASED BY OUR PAST

Perhaps every reader realizes that our past experiences influence our present perceptions. In Chapter 3, we considered shifts in methods and perceptions needed to achieve a TSC. During my workshops on paradigm shifts, someone invariably has expressed concern about resistance. "Those resistors keep playing old tapes and are not open to new ideas," is a common refrain. Past experiences are biasing present perceptions. Actually, there is a long trail of intertwined factors here. Prior experiences filter through a personal evaluation process that is influenced by person factors, including many past perceived experiences. The cumulative collection of those previous experiences biases every new experience and can make it difficult to "teach an old dog new tricks."

Some participants have arrived at my seminars and workshops with a "closed mind" and an "I'm required to be here" attitude. Others start with an "open mind" and an "opportunity to learn" outlook. This is another example of the power of personal perception. How much one learns at those seminars depends on personal perceptions.

Perhaps you will find it worthwhile to copy Figure 5.4 and use it for a group demonstration. You can show how current impressions are affected by prior perceptions by asking participants to call out what they see as you reveal each drawing. The drawings should be uncovered in a particular order. Show the top row of pictures first, revealing each successive picture from left to right. The last picture will probably be identified as the face of an elderly man.

Then, with the top row covered, show the successive animal pictures of the second row. Now, the last picture will likely be identified as a rat or a mouse. Even after knowing the purpose of the demonstration, you can view serially the row pictures in Figure 5.4 and see how your perception of the last drawing changes depending on whether you previously looked at human faces or animals.

FIGURE 5.4 Prior perceptual experience influences current perception. (Adapted from Bugelski & Alimpay, 1961.)

FIGURE 5.5 Viewing this face from a different orientation (by turning the book upside-down) will influence a different perception.

Now, take a look at the woman in Figure 5.5. Do you notice anything strange, other than the picture is upside down? Is this face relatively attractive, or at least normal? Now turn your book upside down and view the woman's face from the normal orientation. Has your perception changed? Why did you not notice her awkward (actually ugly) mouth when the picture was upside down? Perhaps both context and prior experience (or learning) biased your initial perception. I bet this perceptual bias will persist even after you realize the cause of the distortion, and after viewing the face several times in both positions. A biased perception can be difficult to correct. It is not easy to fight human nature.

5.1.3 Relevance to Achieving a TSC

Is the relevance of this discussion to OSH obvious? Perhaps by understanding factors that lead to diverse perceptions, we can become more tolerant of individuals who do not appear to share our opinion or viewpoint. Perhaps the person factors discussed here increase your appreciation and respect for diversity, and support the basic need to actively listen to others with empathy. "Seek first to understand, before being understood" is Stephen Covey's fifth habit for highly effective people (Covey, 1989, p. 235).

It is also possible that this discussion and the exercises on personal perception have reduced your tendency to blame individuals for an injury or to look for a single "root cause" of an undesirable and unintentional incident. Before we react to an incident or injury with our own viewpoint, recommendation, or corrective action, we need to ask others about their perceptions.

I hope I have not reduced your optimism toward achieving a TSC. Perhaps I have alerted you to challenges not previously considered. If I have not convinced you yet to stop claiming "All injuries are preventable," the next section should do the trick.

5.2 PERCEIVED RISK

People are generally underwhelmed or unimpressed by risks or safety hazards at work. Why? Our experiences on the job lead us to perceive a relatively low level of risk. This seems strange. After all, it is quite probable someone will eventually be hurt on the job when you factor in the number of hours workers are exposed to various hazards.

Sensation, Perception, and Perceived Risk

Chapter 4 revealed one major reason for low perceptions of risk in the workplace. It is fundamental, really—we usually get away with risky behavior. As each day passes without receiving an injury, or even a close call, we become more accepting of the common belief, "It's not going to happen to me." Now, let's explore further why we are generally not impressed by safety-related hazards at work.

5.2.1 Real vs. Perceived Risk

Estimating the risk of injury from working with certain equipment is difficult to determine, because work situations vary so dramatically. Plus, the risk can be eliminated completely by the use of appropriate protective clothing or equipment. Still, many people do not appreciate the value of using PPE or following safe operating procedures. Their perception of risk is generally much lower than actual risk. Such thinking pervades society.

Automobile crashes are the nation's leading cause of lost productivity—greater than AIDS, cancer, and heart disease—but how many of us take driving for granted? The risk of a fatality from driving a vehicle or working in a factory is much higher than from the environmental contamination of radiation, asbestos, or industrial chemicals. Yet, recall all of those protests over asbestos in schools and neighborhood chemical plants.

Researchers of risk communication have found that various characteristics of a hazard, irrelevant to actual risk, influence people's perceptions. It is important to consider these characteristics, because behavior is determined by perceived rather than actual risk.

Figure 5.6 lists factors that influence our risk perceptions, as derived from research by Drs. Peter Sandman (1991) and Paul Slovic (1991) and their colleagues. The factors listed on the left reduce perceptions of risk and are typically associated with the workplace. The opposing factors in the right-hand column have been found to increase risk perception, and those are not usually experienced in the work setting. As a consequence, our perception of risk on the job is not as high as it should be, and therefore, we do not work as defensively as we should. Discussing some of these factors will reveal strategies for increasing our own and others' perception of risk in various situations.

Lower Risk	Higher Risk
• exposure is voluntary	• exposure is mandatory
• hazard is familiar	• hazard is unusual
• hazard is forgettable	• hazard is memorable
• hazard is cumulative	• hazard is catastrophic
• collective statistics	• individual statistics
• hazard is understood	• hazard is unknown
• hazard is controllable	• hazard is uncontrollable
• hazard affects anyone	• hazard affects vulnerable people
• preventable	• only reducible
• consequential	• inconsequential

FIGURE 5.6 Factors on the left reduce risk perception and are generally associated with the workplace. (Adapted from Sandman, 1991.)

5.2.2 THE POWER OF PERCEIVED CHOICE

Hazards we choose to experience (like driving, skiing, and working) are seen as less risky than those we feel forced to endure (like food preservatives, environmental pollution, and earthquakes). Of course, the perception of choice is also subjective, varying dramatically among individuals. For example, people who feel they have the freedom to pull up stakes and move whenever they want would likely perceive relatively less risk from a nearby nuclear plant or seismic fault. Likewise, employees who feel they have their pick of places to work generally perceive less risk in their work environment. They are typically more motivated and less distressed. The next chapter reveals evidence-based relations between perceived choice, stress, and distress.

5.2.3 FAMILIARITY BREEDS COMPLACENCY

Familiarity is probably a more powerful determinant of perceived risk than perceived choice. The more we know about a risk, the less it threatens us. Remember how attentive you were when first learning to drive, or when you were first introduced to the equipment in your workplace? It was not long before you lowered your perceptions of risk and changed your behavior accordingly. When driving, for example, most of us quickly shifted from two hands on the wheel and no distractions to steering with one hand while turning up the radio and carrying on a conversation. Then there are those drivers who read and/or send text messages on their cell phone while driving.

5.2.4 SYMPATHY FOR VICTIMS

Many people feel sympathy for the victims of a publicized incident, even vividly visualizing the injury as if it had happened to them. Personalizing those experiences increases perceived risk. At work, employees show much more attention and concern for hazards when injuries or close calls are discussed by the coworkers who experienced them compared to a presentation of statistics. The average person cannot relate to group numbers, but there is power in personal stories. I have met many people over the years who accepted individual accounts in lieu of convincing statistics—"The police officer told Uncle Jake he would have been killed if he had been buckled up"; "Aunt Martha is 91 years old and still smokes two packs of cigarettes a day."

This suggests that we should shift the focus of safety meetings away from statistics and emphasize instead the human element of safety. Safety talks and intervention strategies should center on individual experiences rather than numbers. However, this might be easier said than done. Encouraging victims to come forward with their stories is often stifled by management systems in many companies that seem to value fault-finding over fact-finding, piecemeal rather than system approaches to OSH, and enforcement more than recognition to influence on-the-job behavior.

For ten consecutive keynote presentations at the annual National Safety Council Congress and Expo, I teamed up with Charlie Morecraft, who made OSH personal by revealing his serious workplace injury. With powerful passion and emotion, Charlie spoke first for an hour and explained the severe and long-term losses he experienced following a tragic incident at an Exxon site on August 8, 1980 (8/8/80). He ended each of his dynamic presentations with the assertion that now the audience of safety leaders knows why they do what they do to prevent workplace injuries. Then he introduced me by asserting, "Now Dr. Geller will explain how you can prevent the kind of losses that I have endured."

I always began my contribution by acknowledging the emotions everyone had just experienced from Charlie's story. "Now that you are motivated to prevent injuries, let's talk about what you can do as a safety leader to make sure no one else experiences the physical losses and the psychological distress endured by Charlie Morecraft." Then I had their undivided attention for teaching them evidence-based strategies they could apply to cultivate and achieve a TSC.

Sensation, Perception, and Perceived Risk

5.2.5 Understood and Controllable Hazards

Hazards we can explain and control cause much less alarm than do hazards that are not understood and are therefore perceived as uncontrollable. This reveals a problem with many safety education and training programs. Workplace hazards are explained in a way that creates the impression they can be controlled. Indeed, safety professionals often state a goal of "zero injuries," implying complete control over the factors that cause injuries. This can actually lower perceived risk by convincing people that the causes of occupational injuries are understood and controllable.

Perhaps it would be better for safety leaders to admit and publicize that only two of the three types of factors contributing to workplace injuries can be managed effectively—environmental/equipment factors and work behaviors. As already discussed in preceding chapters, the mysterious inside, unobservable, and subjective world of people dramatically influences the risk of personal injury. Those attitudes, expectancies, perceptions, and personality characteristics cannot be measured, managed, or controlled reliably. Internal human factors make it impossible to prevent all injuries. By discussing the complexity of people and their integral contribution to most workplace hazards and injuries, you can increase both the perceived value of ongoing OSH interventions and the belief that a TSC requires total commitment and engagement from every participant in that culture.

5.2.6 Acceptable Consequences

We are less likely to feel threatened by risk taking or a risk exposure that has its own rewards. However, if few benefits are perceived by an at-risk behavior or an environmental hazard, outrage—or heightened perceived risk—is likely to be the reaction, along with a concerted effort to prevent or curtail that risk.

Some people, for example, perceive guns, cigarettes, and alcohol as having limited benefit and thus lobby to restrict or eliminate those societal hazards. However, the availability of and exposure to those hazards will continue as long as a significant number of individuals perceive that the risk benefits outweigh the risk costs. Cost-benefit analyses are subjective and vary widely as a function of individual experience. For example, the two women in Figure 5.7 obviously perceive the consequences of smoking very differently.

FIGURE 5.7 The perceived consequences of at-risk behavior can vary widely from one person to another.

On the other hand, the benefits of at-risk work behaviors are generally obvious to everyone. For example, it is cooler and more comfortable to work without a respirator. It is also convenient and enables a worker to be more productive. The costs of not wearing the mask might be abstract and delayed if the exposure is not immediately life threatening. Statistics might indicate a relatively low probability of getting a lung disease, which would not surface for decades, if ever. Decisions about risk taking are made every day by workers. By playing the odds and shooting for short-term gains, risky work practices are often accepted and not perceived to be as dangerous as they really are.

5.2.7 Sense of Fairness

Most people believe in a just and fair world. "What goes around comes around." "There's a reason for everything." "People generally get what they deserve." When people receive benefits like increased productivity from their at-risk behavior, the outrage, public attention, or perceived risk is relatively low. On the other hand, when hazards or injuries seem unfair, as when a child is molested or inflicted with a deadly disease, special attention is given. This increased attention results in more perceived risk. This makes it relatively easy to obtain contributions or voluntary assistance for programs that target vulnerable populations, like learning-disabled children. The victims of workplace injuries, however, are not perceived as weak and defenseless. Occupational injuries are indiscriminately distributed among employees who take risks, and "Those people deserve what they get." This is a common perception or attitude, and it lowers the outrage we feel when someone gets injured on the job. Less outrage translates into lower perceived risk.

5.2.8 Risk Compensation

A discussion of risk perception would not be complete without examining one of the most controversial and critical concepts in OSH. Over the years, this phenomenon has been given different labels, including risk homeostasis, risk or danger compensation, risk-offsetting behavior, and perverse compensation. Whatever the label, the basic idea is quite simple and straightforward.

People are presumed to adjust their behavior to compensate for changes in their perception of risk. If a job is made safer with machine guards or the use of PPE, workers reduce their perception of risk and perform more recklessly. Just like thermostats have a set temperature and activate when the temperature deviates from normal, individuals maintain a target level of risk by adjusting their safe versus at-risk behavior in order to achieve a balance between potential risks and perceived benefits.

Figure 5.8 depicts this phenomenon in a common workplace situation. If the use of a back-belt leads to employees lifting heavier loads, then the potential protection from this device has been offset by greater risk taking. The protective device could give a false sense of security and reduce one's perception of being vulnerable to a back injury. The result could be more frequent and heavier lifting and greater probability of an injury. This is why back-belt suppliers emphasize the need for education and training in the proper use of back belts.

Hasanzadeh et al. (2020) immersed participants in a virtual-reality scenario that simulated a roofing task. Then they monitored the workers' reactionary behavior and physiological responses while the participants completed assigned roofing tasks under various levels of safety protection. When providing more safety interventions for these construction workers—by protecting the roof edge with a guard rail and providing the roofer with a fall-arrest system—the roofers increased their risk-taking behavior by as much as 55 percent. The fall-protection devices resulted in the participants stepping closer to the edge of a 20-foot-high virtual roof, leaning over the roof edge, and spending more time exposing themselves to the risk of falling.

Sensation, Perception, and Perceived Risk

FIGURE 5.8 Back belts can give a false sense of security.

5.2.9 Implications of Risk Compensation

What does risk compensation mean for injury prevention? Professor Wilde (1994) says it means safety excellence cannot be achieved through top-down rules and enforcement. Some people only follow the rules when they are supervised and might take greater risks when they can get away with it. As the title of his book *Target Risk* indicates, Dr. Wilde advocates that safety interventions need to lower the level of risk people are willing to tolerate. This requires a change in values.

Dr. Wilde's position is consistent with the theme of this text. When people understand and accept the paradigm shifts needed to achieve a TSC (see Chapter 3), they are on track to reducing their tolerance for risk. Next, they need to believe in the vision of a TSC and buy into the mission of cultivating a TSC. Then, they need to understand and accept the procedures that can contribute to achieving this vision. These methods are explained in Part 3 of this text. Through a continuous process of applying the right procedures, the workforce will feel empowered to actively care for safety. They will come to treat safety as a value rather than a priority, as discussed more fully in Part 4.

5.2.10 Critical Impact of Perceived Choice

The perception of personal choice is a crucial human dynamic to consider here. Ludwig and Geller (1997) found that pizza-delivery drivers demonstrated safer driving when they chose to increase the occurrence of a particular safe-driving behavior (i.e., stopping their vehicles completely before exiting the store's parking lot). Specifically, all pizza delivers stopped completely more often at the target intersection when a safe-stopping goal was set for complete intersection stopping. However, only those drivers who *chose* the intersection-stopping goal were observed to use their safety belts and turn signals more often than did drivers who had been *assigned* to reach the same goal of complete stopping at the intersection.

Relatedly, a behavioral observation study conducted early during the COVID-19 pandemic identified a similar spillover effect or response generalization. Individuals who wore a

COVID-prevention facemask outdoors where facemask wearing was not mandated maintained a greater interpersonal distance from others than did those individuals who were not masked while walking outdoors (Oliver et al., 2021). In this case, as with the pizza-delivery drivers, one safe behavior spilled over or generalized to another safe behavior—the opposite of risk compensation—when a perception of personal choice was implicated.

5.3 IN SUMMARY

This chapter explored the concept of selective sensation or perception and related it to perceived risk and injury prevention. Visual exercises illustrated the impact of past experience and contextual cues on present perception. This enables us to appreciate diversity and realize the value of actively listening during personal interaction. We need to work diligently to understand the perceptions of others before we impulsively jump to conclusions or attempt to exert our influence.

It is important to realize, however, that people often hold on stubbornly to a preconceived notion about someone or something. As illustrated in Figure 5.9, this bias is often caused by prior experience, and it can dramatically affect perception. Perhaps you know this phenomenon as prejudice, one-sidedness, confirmation bias, discrimination, stereotyping, pigheadedness, or just plain bias. Please consider the label "premature cognitive commitment" (Langer, 1989).

I appreciate the term premature cognitive commitment because it makes me mindful of the various ingredients of inflexible prejudice. First, it is premature, meaning it is accomplished before adequate diagnosis, analysis, and consideration. Second, it is cognitive, meaning it is a mental process that influences our perceptions, our attitudes, and our behaviors. Finally, it is a commitment. It is not just a fleeting notion or a temporary opinion. It is a solid, relatively permanent position or sentiment that affects what information a person seeks, attends to, understands, appreciates, believes, and uses.

In fact, premature cognitive commitment is a determinant of much, if not most, interpersonal conflict. It is a barrier we must overcome in order to develop the interdependent teamwork needed

FIGURE 5.9 We all have premature cognitive commitment.

for a TSC. Being mindful of premature cognitive commitment in ourselves and others will not stop this bias, but it is a start.

We must realize that perceptions of risk vary dramatically among individuals. We cannot improve safety unless people increase their perception of, and reduce their tolerance for, risk. Changes in risk perception and risk acceptance will occur when individuals get involved in cultivating a TSC with the principles and procedures discussed in this book.

Several factors were revealed and explained in this chapter that affect whether employees react to workplace hazards with alarm, apathy, or something in between. Taken together, these factors shape personal perceptions of risk and illustrate why the job of improving safety can be so daunting. This justifies more resources for OSH programs, as well as intervention plans to motivate continual employee engagement. Various intervention approaches are introduced and illustrated in Part 4 (Chapters 10–13). However, before discussing strategies for fixing the problem, we need to understand how stress, distress, and personal attributions contribute to the problem. This is our topic for the next chapter.

6 Stress vs. Distress

Stressors can contribute to a close call or an injury; they can be barriers to achieving a TSC. However, stressors can provoke positive stress rather than negative distress, which can lead to constructive problem-solving behavior rather than destructive at-risk behavior. This chapter explains the important distinction between stress and distress and defines factors that determine the occurrence of one over the other.

The concept of "attribution" is introduced as a cognitive process we use to turn stressors into positive stress or negative distress. Attribution bias can reduce distress, but it can also prevent a constructive analysis of an injury or a property-damage incident. This chapter explains the benefits and liabilities of such bias and clarifies its role in shifting stress to distress, or vice versa.

Even if you're on the right track, you'll get run over if you just sit there.

(Will Rogers)

Judy was tired and worried. She had just left her six-year-old son at her sister's house with instructions for her to take him to Dr. Slayton's office for a 10:30 a.m. appointment. She had been up much of the night with Robbie, attempting to comfort him. With tears in his eyes, he had complained of a "hurt" in his stomach. This was the third night his cough had periodically awakened her, but last night Robbie's cough was deeper, seemingly coming from his lungs.

Judy arrived at her workstation a little later than normal and found it more messy than usual. Grumbling under her breath that the night shift had been "careless, sloppy, and thoughtless," she downed her usual cup of coffee and waited for the production line to crank up. She did not clean the work area. After all, it was not her mess. The graveyard shift is not nearly as busy as the day shift. How could they be so sloppy and inconsiderate?

Judy was ready to start her inspection and sorting when she noticed the "load cart" was misaligned. She inserted a wooden handle in the bracket and pulled hard to jerk the cart in place. Suddenly the handle broke, and Judy fell backward against the control panel. Fortunately, she was not hurt, and the only damage was the broken handle. Judy discarded it, inserted another one, and put the cart in place.

During lunch Judy called the doctor's office and learned that her son had the flu, and he would be fine in a day or two. She completed the day in a much better mood, and without incident.

At the end of her shift, Judy filled out a close-call report on her morning mishap. She wrote that someone on the graveyard shift had left her work area in disarray, including a misaligned load tray. She also indicated that the design of the cart handle made damage likely; in the past other cart handles had been broken. She recommended a redesign of the handle brackets, as well as immediate discipline for the graveyard shift in her work area.

The fact that Judy filled out a close-call report is certainly good news, but was that a complete report? Were there some personal factors within Judy that could have influenced the incident? Was Judy under stress or distress, and if so, could that disposition have been a contributing factor? Actually, it has been estimated that from 75 to 85 percent of all industrial injuries can be partially attributed to inappropriate reactions to psychological stress (Jones, 1984).

Judy's close-call report was also clearly biased by common attribution errors researched by social psychologists and used by all of us at times to deflect potential criticism and reduce distress. Attribution errors, along with stress and distress, represent potential barriers to achieving a TSC.

6.1 WHAT IS STRESS?

In simple terms, stress is a psychological and physiological reaction to events or situations in our environment. Whatever triggers the reaction is called a *stressor*. So, stress is the reaction of our mind and body to stressors such as demands, threats, conflicts, frustrations, overloads, competitions, or changes.

Figure 6.1 depicts a scene that might seem familiar—perhaps too familiar. So many people with so much to do and not enough time to do it. Then our goals are thwarted, and our stress turns to distress. Such frustration can lead to aggression and a demeanor that only increases our distress. It is a vicious cycle, and it certainly increases our propensity for personal injury. Certain personality characteristics referred to as "Type A" are more likely to experience the time urgency and competitiveness depicted in Figure 6.1, and these are associated with higher risk for coronary disease when the emotions of anger and hostility are experienced (Dembroski & Costa, 1987; Houston & Vavak, 1991).

It is crucial to distinguish between Type A behavior and Type A emotion. Type A behavior does put people at risk for unintentional injury but not for heart disease. The Type A emotions of anger and hostility, exemplified by "road rage," put people at risk for heart disease and for death following heart disease. Figure 6.2 illustrates survey items that distinguish between Type A behavior and Type A emotion. If you write a number from 1 to 7 next to each item, you can estimate the extent to which each statement applies to you. A "1" would reflect "not at all," a "4" indicates "sometimes," and a "7" would mean "most of the time." The higher your score for the Type A behavior items, the more difficult it is for you to be mindful of momentary risks to your safety. A relatively high score for the Type A emotion items suggest risk of heart disease and a critical need for intrapersonal emotional intelligence, which is explained later in Chapter 15.

FIGURE 6.1 Certain environmental conditions and personality states contribute to stress and distress.

Type A Behavior	Type A Emotion
I experience a lot of time pressure.	People seem to annoy me intentionally.
I feel the pressure to get ahead and succeed.	I often raise my voice.
I do many things fast—talking, walking, eating, and so forth.	Many situations make me angry.
I work long hours.	People consider me short-tempered.
I often want to (and sometimes actually do) finish other people's sentences.	I have a "short fuse" when it comes to tolerating incompetence.
I hate to stand in line.	I try to control my temper, but often lose it.
It is not necessary for others to impose deadlines; I set them for myself.	I express my anger physically by hitting, kicking, slapping, or throwing things.

FIGURE 6.2 Survey items to assess Type A behavior vs. Type B emotion.

6.1.1 Constructive or Destructive?

We usually talk about stress in negative terms, something that is unwanted and uncomfortable. However, the first definition of stress in my copy of American Heritage Dictionary (1985) is "importance, significance, or emphasis placed on something" (page 1205). Similarly, my New Merriam-Webster Dictionary (2023) defines stress as "a factor that induces bodily or mental tension." For a more modern perspective, the generative AI language model ChatGPT defines stress as "as a physiological and psychological response to external or internal pressures or demands, often referred to as stressors. It is a natural reaction that occurs when individuals perceive a situation as challenging, threatening, or demanding, requiring them to adapt or respond."

The undesirable person-state is distress. Distress is defined as "anxiety or suffering . . . severe strain resulting from exhaustion or an accident" (American Heritage Dictionary, 1985, page 410) or "suffering of body or mind: pain, anguish: trouble, misfortune . . . a condition of desperate need" (New Merriam-Webster Dictionary, 2023). Similarly, ChatGPT describes distress as "a state of extreme discomfort, suffering, or emotional anguish. It is a negative emotional response that occurs when an individual feels overwhelmed or unable to cope with a situation or circumstances they are facing. Distress is often associated with feelings of sadness, anxiety, frustration, or despair."

Psychological science supports these distinctions between stress and distress. Stress can be positive, giving us heightened awareness, sharpened mental alertness, and an increased readiness to perform. Certain psychological theories presume that some stress is necessary for people to perform. The person who asserts, "I work best under pressure," understands the motivational power of stress. But can too much pressure, too many deadlines, be destructive?

I am sure most of you have been in situations—or predicaments—where the pressure to perform seemed overwhelming. This is the point where too much pressure can hinder performance, when stress becomes distress. The relationship between external stimulation or pressure to perform and actual performance is depicted in Figure 6.3. This inverted U-shaped function is known as the Yerkes-Dodson law (Yerkes & Dodson, 1908).

Stress vs. Distress

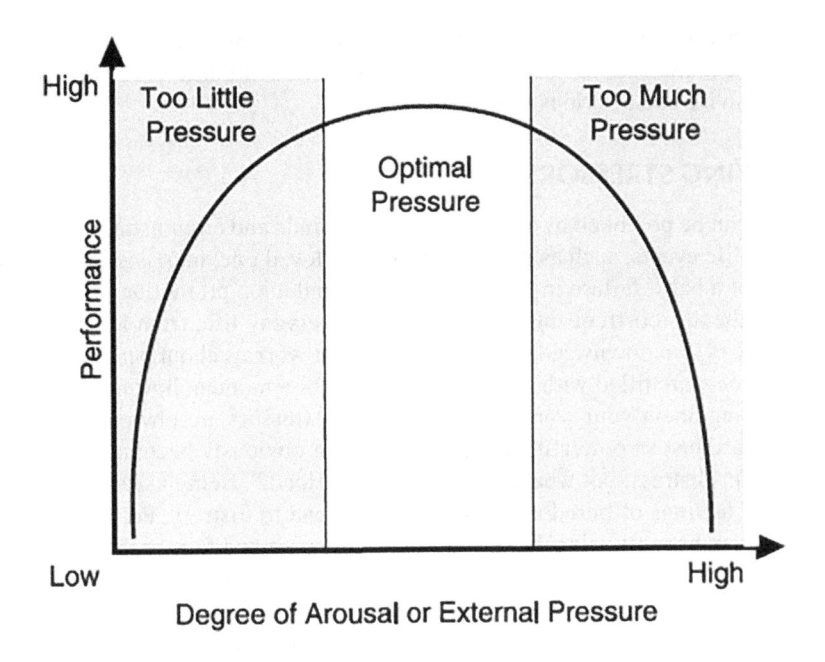

FIGURE 6.3 Arousal from external pressure (or a stressor) improves performance to an optimal point.

The Yerkes-Dodson law states that, up to a point, performance will increase as arousal or pressure to perform well increases. However, the best performance comes when arousal is optimum rather than maximum. Push people too far and their performance starts to deteriorate. In fact, at exceptionally high levels of pressure or tension, an individual might perform as poorly as when they are hardly stimulated at all. Ask someone who is hysterical or someone who is about to fall asleep to do the same job, and you will not be pleased by the results of either individual.

Hans Selye, the Austrian-born founder of stress research, said, "Complete freedom from stress is death" (1974). It is extreme, disorganizing stress we need to avoid, which is distress.

6.1.2 THE EYES OF THE BEHOLDER

Perceptions play an important role in stress and distress. The boss gives a group of employees a deadline; some tighten up inside, others take it in stride. Some circle their calendars and cannot take their minds off the due date. Others seem to pay it no mind.

When a stressor is noticed and causes a reaction, the result can be constructive or destructive. If we believe we are in control—that we can deal with the overload, frustrations, conflicts, or whatever is triggered by the stressor—we become aroused and motivated to go beyond the call of duty. We actually achieve more. On the other hand, when we believe we cannot handle the demands of the stressor, the resulting psychological and physiological reactions are likely to be detrimental to our performance, as well as our health and safety.

Why does a manager's deadline motivate one person and distress another? It depends on a number of internal person factors. These include the amount of arousal already present in the individual and the person's degree of preparedness or self-confidence.

Those "butterflies" we feel in our stomach can help or hinder performance, depending on personal perception. When the "butterflies" are aligned for goal-directed behavior, we feel in control of the situation, prepared for it, and challenged by it. The stress is positive—it arouses or motivates performance improvement. When the "butterflies" are misaligned and scattered in different directions, we feel unprepared, burdened rather than challenged. The stressor is experienced as distress. This arousal can divert our attention, interfere with thinking processes, disrupt performance, and reduce our ability and overall motivation to perform well. The result can be at-risk behavior and a serious injury.

6.2 IDENTIFYING STRESSORS

Stress or distress can be provoked by a wide range of demands and circumstances. Some stressors are acute, sudden life events, such as death or injury to a loved one, marriage, marital separation or divorce, birth of a baby, failure in school or at work, and a job promotion or relocation. Other stressors include the all-too-frequent minor hassles of everyday life, from long lines and excessive traffic (Figure 6.1) to downsized work conditions and worries about personal finances.

Our jobs or careers are filled with stressors. Consider for a moment how much time you spend working, or thinking about your work. Some workplace stressors are obvious; others might not be as evident but are just as powerful. Work overload can obviously become a stressor and provoke either stress or distress, but what about "work *under*load?" Being asked to do too little can produce profound feelings of boredom, which can also lead to distress. Performance appraisals are stressors that can be motivational if perceived as objective and fair, or they can contribute to distress and inferior performance if viewed as subjective and unfair.

Other work-related factors that can be perceived as stressors and lead to distress and eventual burnout include: role conflict or ambiguity; uncertainty about one's job responsibilities; responsibility for others; a crowded, noisy, smelly, or dirty work environment; lack of involvement or participation in decision-making; interpersonal conflict with other employees; and insufficient support from coworkers.

Let's review the key points about stress and distress. Actually, the flow schematic in Figure 6.4 says it all. First, an environmental event is perceived and appraised as a stressor to be concerned about or as a harmless or irrelevant stimulus. An event is perceived as a stressor if it involves *harm* or loss that has already occurred, a *threat* of some future danger, or a *challenge* to be overcome.

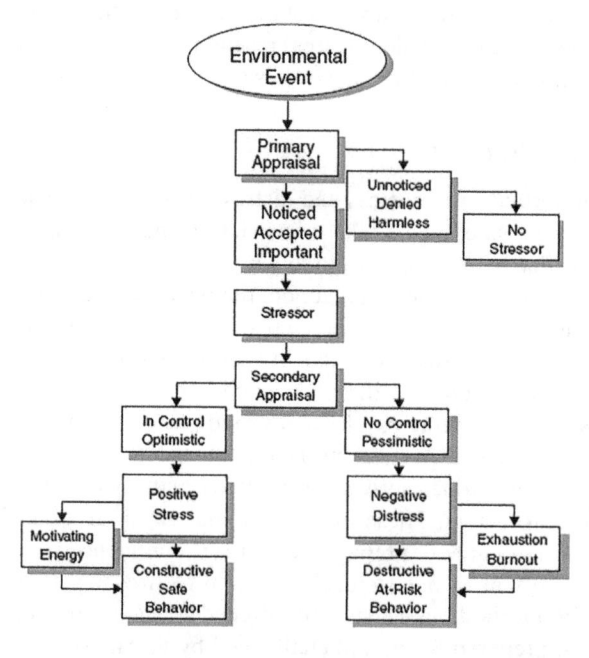

FIGURE 6.4 Through personal appraisal, people transform stressors into positive stress or negative distress.

Stress vs. Distress

Harm is how we appraise the impact of an event. For example, if you oversleep and miss an important safety meeting, the damage is done. In contrast, *threat* is how we assess potential future harm from the event. Missing the safety meeting could lower your team's opinion of you and reduce your opportunity to get actively involved in a new safety process. *Challenge* is our appraisal of how well we can eventually profit from the damage done. You could view missing the safety meeting as an opportunity to learn from one-on-one discussions with coworkers later. That could demonstrate your personal commitment to OSH and enable you to collect diverse opinions.

In that latter case, you are perceiving the stressor as an opportunity to learn and show commitment. Such an evaluation occurs during the secondary appraisal stage and the result can be positive and constructive. On the other hand, your appraisal could be downbeat—you see no recourse for missing the safety meeting, and so you do nothing about it. Now, coworkers might think you do not care about the new OSH process and express disappointment. In turn, you might give up or actively resist participating. That outcome would be nonproductive, of course, and possibly destructive.

The secondary appraisal stage, as depicted in Figure 6.4, determines whether the stressor leads to positive stress and constructive behavior or to negative distress and destructive behavior. The difference rests with the individual. Does s/he assess the stressful situation as controllable and, thus, remains optimistic during attempts to cope with the stressor?

As illustrated in Figure 6.5, when people judge their stressors as uncontrollable and unmanageable, a helpless or pessimistic attitude can prevail and lead to distress and destructive or even at-risk behavior. Several personal, interpersonal and environmental factors influence whether this secondary appraisal leads to constructive or destructive behavior. This is the theme of the next section.

FIGURE 6.5 Lack of perceived control can lead to distress.

6.3 COPING WITH STRESSORS

Understanding the multiple causes of conflict, frustration, overload, boredom, and other potential stressors in our lives can sometimes lead to effective coping strategies. These include:

- Revising schedules to avoid hassles like traffic and shopping lines.
- Refusing a request that will overload you.
- Finding time to truly relax and recuperate from tension and fatigue.
- Communicating effectively with others to clarify work duties, reduce conflict, gain support, or feel more comfortable about added job responsibilities.
- Getting reassigned to a task that better fits your present talents and aspirations.

Unfortunately, it is often impossible to avoid sudden (acute) or continual (chronic) stressors in our lives. We need to deal with these head-on. Believing you can handle the harm, threat, or challenge of stressors is the first step toward experiencing stress rather than distress, and acting constructively rather than destructively.

6.3.1 Person Factors

Certain personality characteristics make some people more resistant to distress. Individuals who believe they control their own destinies and generally expect the best from life are, in fact, more likely to gain control of their stressors and experience positive stress rather than negative distress. It is important to realize that such person-states (e.g., self-mastery and optimism) are not permanent inborn traits of people. They are states of mind or expectations derived from personal experience, and they can be nurtured. It is possible to give people experiences that increase feelings of being "in control"—experiences that lead individuals to believe something good will come from their attempts to turn stress into constructive action.

Learning to feel helpless. When I helped clients assess the safety climate of their workplaces, I often uncovered an attitude among hourly workers, and some managers as well, that exemplified an important psychological concept called "learned helplessness." For instance, when I asked workers what they do regularly to make their workplace safer, I often heard:

"Besides following the safety procedures there's not much I can do for safety around here."
"It is what it is."
"It really doesn't matter much what I do; whatever will be will be."
"There's not much I can do about reducing work injuries; if it's my time, it's my time."

Those statements reflect "learned helplessness." That concept was labeled almost 50 years ago by research psychologists studying the learning process of dogs (Seligman, 1975). They measured the speed at which dogs learned to jump a low barrier separating two chambers in order to avoid receiving an electric shock through the grid floor. A tone or light would be activated, then a shock was applied to the grid floor of the chamber. At first, the dogs did not jump the hurdle until receiving the shock, but after a few trials the dogs learned to avoid the shock by jumping into the other chamber as soon as the warning signal was presented.

Some dogs experienced shocks regardless of their behavior before the regular shock-avoidance learning trials. Those dogs did not learn to jump the barrier to escape the shock. Instead, they typically just laid down in the shock chamber and whimpered. The earlier bad experience with inescapable shock had taught these dogs to be helpless. Dr. Seligman and associates coined the term "learned helplessness" to describe this state. Their finding was subsequently demonstrated in a variety of follow-up experiments with human subjects.

Note how prior failures conditioned research subjects—ranging from dogs to humans—to feel helpless, in fact to be helpless. It is rather easy to assume that workers develop a "helpless" perspective regarding OSH as a result of bad prior experiences. If their safety suggestions are ignored, or policies and procedures always come from management, workers might learn to feel helpless about OSH. It is also true, however, that life experiences beyond the workplace can shape an attitude of learned helplessness. Some individuals come to work with a greater propensity to feel helpless in general, and this can spill over to feelings regarding OSH.

Stress vs. Distress

FIGURE 6.6 Perception affects expectation which affects behavior, which in turn affects perception.

Learned optimism. A bad experience does not necessarily lead to an attitude of learned helplessness. You probably know people who seem to derive strength or energy from their failures, and they try even harder to succeed when given another chance. Similarly, Seligman and colleagues found that certain dogs resisted learned helplessness if they previously had success avoiding the electric shock. Likewise, some people tend to give up in the face of a stressor, while others fight back.

You probably recognize this difference between learned helplessness and learned optimism as the more popular pessimist vs. optimist distinction. As you have heard it asked before, "How do you see the glass of water?" Is it half full or half empty? We see it differently, depending on our current person-state of optimism or pessimism. This contrast in personal perception is illustrated humorously in Figure 6.6. The point is that our personality, past experience, and current situation influence whether we feel optimistic and in control or pessimistic and out of control.

What can be done to help those who feel helpless? How can we get them to commit to and participate in the proactive processes of injury prevention? The work culture can play a critical role here. Actively caring for safety happens when employees feel empowered to make a difference and perceive they are successful.

When workers believe through personal experience that their efforts can make a difference in OSH, they develop an antidote for learned helplessness. This has been termed "learned optimism." If the corporate climate empowers workers to take control and manage safety for themselves and their coworkers, they can legitimately attribute safety success to their own actions. This enhances learned optimism and feelings of being in control. Besides seeing the glass as half full, optimistic people under stress find ways to fill the rest of the glass.

6.3.2 Fit for Stressors

Fitness is another way to increase our sense of personal control and optimism. Being physically fit increases our body's ability to cope with stressors. You probably know the basic guidelines for improving fitness, which include stopping smoking; reducing or eliminating alcohol consumption; exercising regularly, at least three times a week for about 30 minutes per session; eating balanced meals—especially breakfast—with decreased fat, salt, and sugar; and obtaining enough sleep (usually 7 or 8 hours per 24-hour period for most people). Some of us find that following those guidelines over the long haul is easier said than done. We need support and encouragement to break a smoking or drinking habit or to maintain a regular exercise routine.

Figure 6.7 illustrates the type of behavior that accompanied the computer revolution. Low physical activity has become the way of work life for many of us. Often this inactivity spills over into home life. Survey research has shown that only one in four Americans exercises regularly and intensely enough to reduce the risk of stressor-induced heart disease (U.S. Department of Labor, 2023). Figure 6.7 also depicts smoking behavior, considered the largest preventable cause of illness and premature death in the U.S., and accounting for approximately 480,000 deaths in 2021 (Centers for Disease Control and Prevention, 2023).

Also portrayed in Figure 6.7 is the positive influence of perceived choice and personal control. Although the behavior is essentially the same at work (10 to 5) and at home (5 to 10), the individual is seemingly much happier at home. Why? Because at home he holds the remote control and therefore perceives more personal choice and control.

FIGURE 6.7 Becoming a "mouse potato" by day and a "couch potato" by night can reduce one's physical ability to cope with stressors.

However, personal choice/control is truly in the eyes of the beholder. Figure 6.8 depicts legitimate perceptions of choice/control from the subjects of an experiment. Those rodents are not usually considered "in control" of the situation, but in many ways they are. By simply changing our perspective, we can often perceive personal choice and accept more personal control at work, and such a change in mindset can turn negative distress into positive stress.

6.3.3 Perceived Choice

The freedom to choose is invaluable for human wellbeing and life satisfaction. Yet, we often take this crucial human dynamic for granted, without realizing the dramatic influence of perceived choice. Consider, for example, the extent to which the pleasure of a weekend break from a weekly work routine is determined by the opportunity to choose what to do each day and when to do it—from choosing when to wake in the morning to deciding your daily activities, some of which might require more energy and effort than that required to accomplish typical work assignments.

MAN, DO WE HAVE THIS
GUY CONTROLLED. EVERY TIME
WE PULL THE LEVER HE GIVES US
A FOOD PELLET.

FIGURE 6.8 Even the most obvious top-down situation allows for perceptions of bottom-up control.

In other words, your greater enjoyment performing one task over another is often due to the perception of having more personal choice for that task than the other.

Participative management. Many readers are familiar with the term "participative management," which essentially means employees are provided with some personal choice during the planning, execution, and/or the evaluation of their jobs. The result: more self-motivation, engagement, and life satisfaction throughout the workforce and less distress. However, participative management takes more time and is therefore less efficient than a management-directed approach, but the gain in effectiveness is typically well worth the loss of efficiency.

Consider two approaches to soliciting OSH assistance from a work team. A common approach is for the supervisor to define both the OSH issue and the solution and then to assign certain employees relevant problem-solving tasks. Alternatively, the supervisor might solicit volunteers to perform the various problem-solving tasks and thereby integrate some perceived choice into the scenario.

A more effective approach would be to describe the OSH issue to the work team and ask the workers for their solution possibilities, including the various problem-solving tasks needed. Obviously, this approach will require much more interpersonal dialogue, and therefore it will be substantially less efficient than the first approach. However, the work team might derive a more effective solution. But, even if the problem-solving tasks derived by the employees are the same as those proposed by the supervisor using the first approach, the perceived choice and self-motivation inspired by the second "less efficient" approach would be well worth the extra time and effort.

Activating cognitive dissonance. People need to feel congruence or consistency between what they think, what they feel, and what they do. We experience tension or cognitive dissonance (Festinger, 1957) when perceiving an inconsistency between a behavior we choose to perform and another behavior, cognition, or personal value. The key word in that sentence is choose. When we feel coerced or incentivized into doing something, we do not feel uncomfortable if that behavior is inconsistent with a personal attitude, value, or another behavior. Can an OSH intervention activate cognitive dissonance and thereby increase personal responsibility for injury prevention? Let's explore that question, which includes the critical relevance of perceived choice.

Recognizing an inconsistency between an internal conviction and self-directed behavior causes mental tension and self-motivation to restore congruity between a behavior and one's belief, attitude, or value. Simply put, we want our actions to reflect our values, and vice versa. When an inconsistency between a value and a self-directed behavior is experienced, behavior is typically adjusted to match the value rather than changing a value to match the behavior.

An intervention process to activate cognitive dissonance and safety-related behavior is simple and straightforward. Inspire people to declare safety a value linked to the changing priorities of every workday. Then, specify behaviors that are compatible versus incompatible with that value statement. Cognitive dissonance can be activated whenever someone points out a behavior that is inconsistent with safety as a core value.

For example, after observing an at-risk work behavior, remind the performer of the group consensus that safety is considered a value at their workplace. If the individual realizes the inconsistency, that worker should proceed to resolve the cognitive dissonance or behavior-value imbalance by substituting safe behavior for the observed at-risk behavior. However, please note the fundamental influence of perceived choice here. The individual's behavioral conviction should not be viewed as controlled by extrinsic contingencies, like incentives, disincentives, or peer pressure, but by a self-motivated decision to demonstrate safety as a core value. In other words, cognitive dissonance is only experienced when the behavior-value inconsistency is perceived as voluntary or self-directed.

More than two decades ago, I described the "hypocrisy effect" as an intervention to motivate the occurrence of safety-related behavior by having participants experience a discrepancy between what they have done in the past and what they should do (Geller, 2000). This intervention process is as follows: 1) present the rationale for a particular safe behavior, 2) ask the participants to make a commitment to always choose this safe behavior over designated at-risk practices, and then 3) ask the participants to list the most recent times they had performed an at-risk alternative to the safe behavior.

Note how this hypocrisy-based intervention activates cognitive dissonance. The objective is to provoke the participants into experiencing a discrepancy between their behavior and a commitment or personal-value statement. This triggers cognitive dissonance, which ends when self-directed action restores the imbalance.

The more public the commitment or value affirmation, the greater its impact. Thus, when people attest to safety as a value in the presence of others, they feel a special sense of obligation to live up to their affirmation. Those who hear someone declare safety as a value can readily activate cognitive dissonance within that individual by calling attention to a particular at-risk behavior they observe from that person that does not reflect safety as a value.

6.3.4 Social Factors

A support system of friends, family, and coworkers can do wonders at helping us reduce distress in our lives. Social support can motivate us to do what it takes to stay physically fit, and the people around us can make a boring task bearable and even satisfying. Of course, they can also turn a stimulating job into something dull and tedious. It works both ways. People can motivate us or trigger conflict, frustration, hostility, a win/lose perspective, and distress. It is up to us to make the most of the people around us. We can learn from those who take effective control of stressful situations and expect the best, or we can listen to the complaining, backstabbing, and cynicism of others and fuel our own potential for distress.

It is obviously important to interact with those who can help us build resistance against distress and help us feel better about potential stressors. We can also set the right example and demonstrate the kind of social support for others that we want for ourselves. The good feelings of personal control and optimism you experience from reaching out to help others can do wonders in helping you cope with your own stressors. This actively caring for people (AC4P) stance builds your own support system, which you might need if your own stressors get too overwhelming to handle yourself.

The next section of this chapter introduces another means of reducing distress. It is a phenomenon that has particular implications for OSH. In the aftermath of an injury or close call, this human dynamic can distort injury reports and analyses, resulting in inappropriate or less-than-optimal suggestions for corrective action. This phenomenon of attributional bias can also create communication barriers between people and limit the cooperative participation needed to achieve a TSC.

6.4 ATTRIBUTIONAL BIAS

Think back to the anecdote at the start of this chapter. It was suggested that Judy's close-call report was incomplete or biased. Specifically, Judy did not report the potential influence of her own distress on the incident. Rather, she focused on factors beyond her immediate control—the poor bracket design for the wooden handle and the messy work area left by others. Giving up personal responsibility eliminated the incident as a stressor for her. She did not have to deal with any guilt for almost hurting herself and damaging property. Her denial eased her distress but biased the close-call report. Psychologists refer to this as *attributional bias*. By understanding when and how this human dynamic occurs, we can focus injury analysis on finding facts rather than faults. This is a paradigm shift needed to achieve a TSC.

6.4.1 THE FUNDAMENTAL ATTRIBUTION ERROR

Every day, we struggle to explain the actions of others. Why did she say that to me? Why did the job applicant refuse to answer that question? Why did Joe leave his work station in such a mess? Why did the secretary hang up on me? Why did Gayle take that sick leave? Why does she allow her young children to ride in the bed of her pick-up truck? Why did the motorist pull a gun out of his glove compartment to shoot someone in the next car? Why do tragic mass shootings at schools and elsewhere occur? When trying to answer questions like these, do we point to external environmental factors, such as equipment malfunctioning, excessive traffic, warm climate, and work demands, or do we focus on internal person factors, such as personality, intelligence, attitude, or frustration?

Social psychologists have discovered a *fundamental attribution error* when systematically studying how people explain the behavior of others. When evaluating others, we tend to overestimate the influence of internal factors and underestimate the role of external factors. We are more apt to judge the job applicant as rude or unaware (internal factors) than caught off guard by a confusing or unclear question (external factor). Joe was sloppy or inconsiderate rather than overwhelmed by production demands. The injured employee was careless rather than distracted by a sudden environmental noise.

That is how we perceive things when we are judging others. However, it is different when we evaluate ourselves. The individuals performing the behaviors in the previous paragraph would say the causes were due more to external than internal factors.

Here's a personal example. My university students are quick to judge me as being an extrovert—outgoing and sociable. When I lecture in large classes of 600 to 800 students, I am animated and passionate, and they attribute my performance to internal personality traits, but I know better. I see myself in many different situations and realize just how much my behavior changes depending on where I am. In many social settings I am downright shy and reserved and very sensitive to external factors.

6.4.2 THE SELF-SERVING BIAS

Students who flunk my university exams are quick to blame external factors, like tricky questions, wrong reading material assigned, and unfair grading. In contrast, students who do well are quite willing to give themselves most of the credit. It was not that I taught them well or that the exam questions were straightforward and fair; rather the student is intelligent, creative, motivated, and prepared. This real-world example, to which I bet most readers can relate, illustrates another type of attributional distortion, referred to as the *self-serving bias*.

How can this bias affect an incident or injury analysis? Consider Judy's close-call experience. She protected her self-esteem by overestimating external causes and underplaying internal factors. People will go to great lengths to avoid blame for unintentional property damage or injury. This reduces negative stress or distress. No one wants to feel responsible for a workplace injury, especially if the company focuses on reducing "the numbers," such as the plant's total recordable injury rate (TRIR).

It is important for us to acknowledge how perceptions can be biased. Outsiders tend to blame the victim; victims look to extenuating circumstances. We should empathize with the self-serving bias of the victim because it will reduce that person's distress. It will shift attention to external factors that can be controlled more readily than internal factors related to a person's attitude, mood, or state of mind.

6.5 IN SUMMARY

This chapter explicated the difference between stress and distress and discussed some strategies for reducing distress or turning negative distress into positive stress. Stress and distress begin with a stressor that can be a major life event or a minor irritation of everyday living. You can evaluate or appraise the stressor in a way that is constructive—resulting in safe behavior—or destructive—influencing at-risk behavior. When people are physically fit, in control, optimistic, and able to rely on the social support of others, they are likely to turn a stressor into constructive energy for achieving success. This is positive stress.

When stressors are perceived as insurmountable and unavoidable, distress is likely. Without adequate support from others, this condition can lead to physical and mental exhaustion, at-risk behavior, and unintentional injury to oneself or others. We need to become aware of the potential stressors in our lives and in the lives of our coworkers. In addition, we need to develop personal and interpersonal strategies to prevent distress in ourselves and others.

The work culture, including policies, paradigms, and personnel, can have a dramatic impact on whether victims of close calls, injuries, or other adversities experience stress or distress. The fundamental attribution error, whereby we overestimate personal factors to explain others' behavior ("Judy broke the handle because she was tired, stressed out, and careless"), can provoke distress and pinpoint the very aspects of an incident most difficult to define and control.

A victim's natural tendency to reveal a self-serving bias when discussing an incident—by putting more emphasis on external, situational causes—should be supported by the work culture. This reduces the victim's distress and puts the focus on the observable factors, including behavior, most readily defined and influenced. Practical procedures for doing this are detailed in the next three chapters—Part 3: Behavior-Based Psychology.

Part Three

Behavior-Based Psychology

7 Basic Principles

To achieve a Total Safety Culture, we need to integrate behavior-based and person-based psychology and effect large-scale culture change. The three chapters in Part 3 explain principles and procedures founded on applied behavioral science (ABS) that can be applied successfully to improve behaviors and attitudes throughout organizations and communities. This chapter describes the primary characteristics of the behavior-based approach for the prevention and the treatment of human problems, and shows special relevance to OSH. The three basic ways we learn are reviewed and related to the development of safe vs. at-risk behaviors and attitudes.

> One can picture a good life by analyzing one's feelings, but one can achieve it only by arranging environmental contingencies.
>
> **(B. F. Skinner)**

Specific safety techniques can be viewed as possible routes to reach a destination, in our case, a TSC. A particular route may be irrelevant or need to be modified substantially for a given work culture. The key is to begin with a complete and accurate map. In other words, it is most important to start with an understanding of the basic principles.

As you might recall, our overall map or guiding principle is represented by the Safety Triad (Figure 2.3). Its reference points are the three primary determinants of OSH-related performance—environment, person, and behavior factors. To achieve a TSC, we need to understand and pay attention to each of these.

In Part 2, a number of person-based factors were addressed that can contribute to injuries, including cognitions, perceptions, and attributions. The BASIC ID acronym was introduced in Chapter 4 to illustrate the complexity of human dynamics and the special challenges involved in preventing injuries. Behavior was the first dimension discussed, and it is implicated directly or indirectly in each of the other dimensions. Attitudes, sensations, imagery, and cognitions—the thinking, person-state side of the Safety Triad—are each influenced by behavior. That is what is meant by the phrase, "Acting people into changing their thinking." When we change our behaviors, such as adopting a new strategy or paradigm, certain person-states also change.

The reverse is also true. Changes in attitudes, sensations, imagery, and cognitions can alter behaviors. However, considerable research has shown that it is easier and more cost effective to "act people into changing their thinking" than the reverse, especially in organizational and community settings.

7.1 PRIMACY OF BEHAVIOR

Whether treating clinical problems (such as drug abuse, sexual dysfunction, depression, anxiety, pain, hypertension, and child or spouse abuse) or preventing any number of health, social, or environmental ills (from developing healthy and safe lifestyles to improving education and protecting the environment), overt behavior is the focus. Treatment or prevention is based on three basic questions:

1. What behaviors need to be increased or decreased in frequency of occurrence to solve or prevent the problem?
2. What environmental conditions, including interpersonal relations, are currently supporting undesirable behaviors or inhibiting desirable behaviors?

3. What environmental or social conditions can be changed to decrease the occurrence of undesirable behaviors and increase the occurrence of desirable behaviors?

Thus, behavior change is both the outcome and the means. It is the desired outcome of treatment or prevention, as well as the means to solving the identified problem.

7.1.1 Reducing At-Risk Behaviors

Heinrich's well-known Law of Safety implicates at-risk behavior as a root cause of most close calls and injuries (Heinrich, 1931; Heinrich et al., 1980). Over the past 40 years, various behavior-based research studies have verified this aspect of Heinrich's law by systematically evaluating the impact of interventions designed to decrease the frequency of employees' at-risk behaviors. Feedback from behavioral observations was a common component in most of the successful intervention processes, whether the behavioral feedback was delivered verbally, graphically by tables and charts, or through corrective action.

The behavior-based approach to preventing injuries is implicated in Figure 7.1. At-risk behaviors are presumed to be a major cause of a series of progressively more serious incidents, from a close call to a fatality. According to Heinrich's law, there are numerous risky acts for every close call and many more close calls than lost-time injuries. This is fortunate news, but let us not forget that timing or luck is often the only difference between a close call and a serious injury.

Typically, behavior-change techniques are applied to specific targets. It is necessary, of course, that participants know why certain targeted behaviors are undesirable and that they have the physical ability to avoid them. Education and engineering interventions are sometimes needed to satisfy the physical and knowledge factors of Figure 7.1. The execution factors represent the motivational aspect of the problem and usually require the most attention. In other words, people

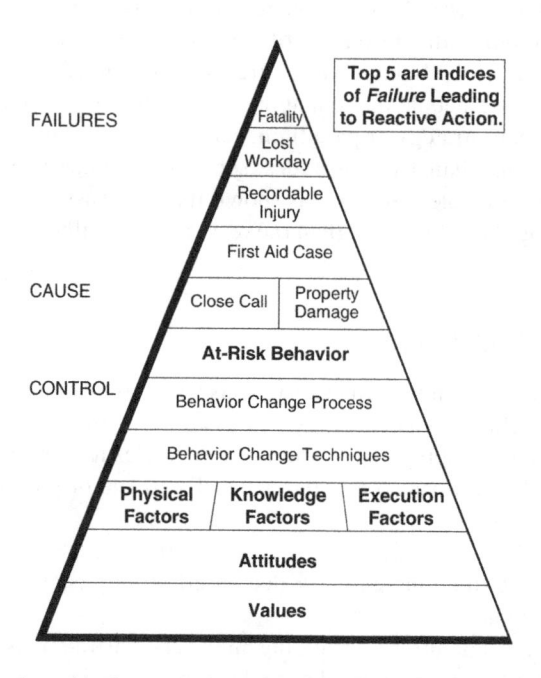

FIGURE 7.1 Behavior-based safety can decrease at-risk behavior in order to avoid failure.

Basic Principles

usually know what at-risk behaviors to avoid and they have the ability to do so, but their motivation might be lacking or misdirected. Behavior-change techniques are used to align individual and group motivation with the avoidance of undesired at-risk behavior.

Values and attitudes form the foundation of the pyramid in Figure 7.1. These obviously critical person factors need to support the safety process. Recall our discussion about risk compensation in Chapter 5 and Dr. Wilde's warning that it is more important to reduce risk tolerance than increase compliance with specific safety rules (1994). This happens when people believe in the safety process and help to make it work. Behavior helps to make the process work, and if involvement is voluntary and appropriately recognized, the process will lead to supportive attitudes and values for keeping the process going.

The behavior-based approach illustrated in Figure 7.1 is failure oriented. It is also more reactive than proactive. The outcome measures are failures—fatalities, lost workdays, and the like—that require a fix.

The reactive and punitive approach is typical for government agencies. The most convenient way to control behavior is to pass a law and enforce it. In fact, as depicted in Figure 7.2, this is the standard government approach to safety improvement. When agents of the Occupational Safety and Health Administration visit a site for inspection, they expect to write citations. They look for mistakes or failures, thereby hoping to improve behavior through the threat of a negative consequence—negative reinforcement. Unfortunately, this perspective can promote negative attitudes about the entire OSH process.

It is usually better to focus on increasing the occurrence of safe behaviors. This is a proactive approach. When safe behaviors are substituted for at-risk behaviors, injuries will be prevented. By emphasizing safe behaviors, employees feel more positive about the process and are more willing to participate.

FIGURE 7.2 A reactive and punitive approach to safety promotes avoiding failure rather than achieving success.

7.1.2 Increasing Safe Behaviors

Figure 7.3 illustrates a positive and proactive behavior-based model. I do not recommend this instead of the corrective action approach depicted in Figure 7.1. A complete behavior-based process should target what is right and what is wrong about a particular work routine, but again, more employees will participate with a positive attitude and remain committed over time if there is more recognition of achievements than correction of mistakes.

Monitoring achievement. The indices of achievement in Figure 7.3 are generally more difficult to record and track than the indices of failure in Figure 7.1. Actually, the failure outcomes in Figure 7.1 are observed and recorded quite naturally. Except for close calls and first-aid cases, the failures in Figure 7.1 have traditionally resulted in systematic investigation and formal reports.

In contrast, the achievements in Figure 7.3 are somewhat difficult to define and record. In fact, it is impossible to obtain an objective record of the number of injuries prevented. However, a reasonable estimate of injuries prevented could be calculated after you achieve a consistent decrease in injuries as a result of a proactive, behavior-based process.

It is possible to derive direct and objective definitions of the other success indices in Figure 7.3, and to use those to estimate overall achievement. Involvement, for example, can be defined by recording participation in voluntary programs, and incidents of corrective action can be counted in a number of situations. You could chart the number of safety work orders turned in and completed, the number of safety audits conducted and safety suggestions given, and the number of safety improvements occurring as a result of safety-related suggestions and close-call reports.

The safety share. The "safety share" noted in Figure 7.3 is a simple behavior-focused process that reflects an emphasis on achievement. At the start of group meetings, the leader simply asks participants to report something they have done for safety during the past week or since the last meeting. Because the "safety share" is used to open all kinds of meetings, safety is given special status and integrated into the overall business agenda. From my experience with this process, people come to expect queries about their safety accomplishments, and many go out of their way to have an impressive safety story to share.

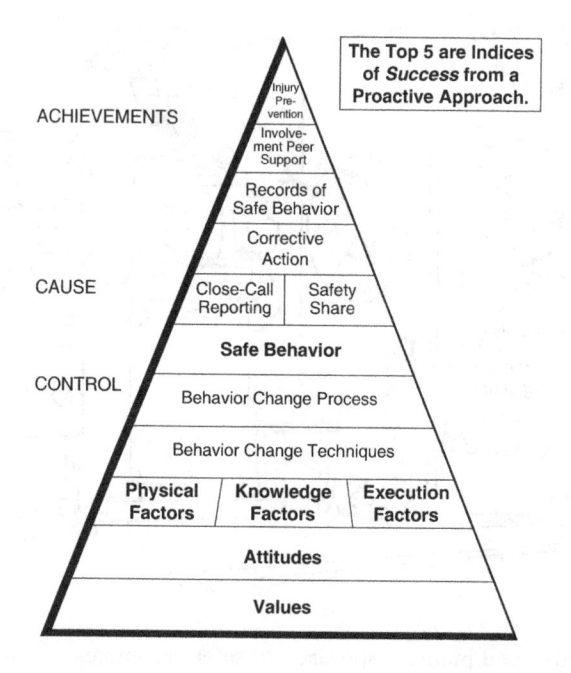

FIGURE 7.3 Behavior-based safety can increase safe behavior in order to achieve success.

Basic Principles

This simple awareness booster—"What have you done for safety?"—helps teach an important lesson. Employees learn that OSH is not only loss control—an attempt to avoid failure—but can be discussed in the same terms of achievement as the quality and quantity of production, and profits. As a measurement tool, it is possible to count and monitor the number of safety shares offered per meeting as an estimate of proactive OSH success in the work culture.

7.2 LEARNING FROM EXPERIENCE

A key assumption of the behavior-based approach is that behavior (both desirable and undesirable) is learned and can be changed by providing people with new learning experiences. Diverse cultural, social, environmental, and biological factors interact to influence our readiness to learn behaviors. These factors also support or hinder behaviors once they are learned. We do not understand exactly how the particulars work—diverse factors interacting to influence behaviors for each individual. However, basic ways to develop, support, and maintain desirable behaviors have been researched for practical real-world application.

Psychologists define learning as a change in behavior, or potential to behave in a certain way, as the result of direct and indirect experience. In other words, we learn from observing and experiencing events and behaviors in our surroundings. While the effects of learning are widespread and varied, it is generally believed that there are three basic learning processes: classical conditioning, operant conditioning, and observational learning.

7.2.1 CLASSICAL CONDITIONING

This form of learning became the subject of careful study in the early 20th century with the seminal research of Ivan Pavlov (1849–1936), a Nobel Prize-winning physiologist from Russia. Pavlov (1927) did not set out to study learning; rather his research focused on the process of digestion in dogs. He was interested in how reflex responses were influenced by stimulating a dog's digestive system with food. During his experimentation, he serendipitously found that his subjects began to salivate before actually tasting the food. They appeared to anticipate the food stimulation by salivating when they saw the food or heard the researchers preparing it. Some dogs even salivated when seeing the empty food pan or the person who brought in the food. Through experiencing the relationship between certain stimuli and food, the dogs learned when to anticipate food. Pavlov recognized this as an important phenomenon and shifted the focus of his research to address it.

The top half of Figure 7.4 depicts the sequence of stimulus–response events occurring in classical conditioning. Actually, before learning occurs, the sequence includes only three events—conditioned stimulus (CS), unconditioned stimulus (UCS), and unconditioned response (UCR). The UCS elicits a UCR automatically, as in an autonomic reflex. That is, the food (UCS) triggered a salivation reflex (UCR) in Pavlov's dogs. In the same way, the smell of popcorn (UCS) might make your mouth water (UCR); a puff of air to your eye (UCS) would result in an automatic eyeblink (UCR), ingesting certain drugs (like Antabuse) would cause a nausea reaction; and a state trooper writing you a speeding ticket is likely to influence an emotional reaction, like distress, nervousness, or anger.

If a particular stimulus (CS) consistently precedes the UCS on a number of occasions, the reflex (or involuntary response) will become elicited by the CS. This is classical conditioning and occurred when Pavlov's "slobbering dogs" salivated when they heard the bell that preceded food delivery. Classical conditioning would also occur if your mouth watered when you heard the bell from the microwave oven tell you the popcorn was ready, if you blinked your eye following the illumination of a dim light that consistently preceded the air puff, if you felt nauseous after seeing and smelling the alcoholic beverage that previously accompanied the ingestion of Antabuse, and if you got nervous and upset after seeing a flashing blue light in your vehicle's rearview mirror.

Classical Conditioning				
Conditioned Stimulus	Conditioned Response (involuntary)		Unconditioned Stimulus	Unconditioned Response (involuntary)
CS —Elicits→	CR	No Contingency	UCS —Elicits→	UCR
Bell	Salivation		Food	Salivation
Blue Light	Emotional Reaction		State Trooper	Emotional Reaction

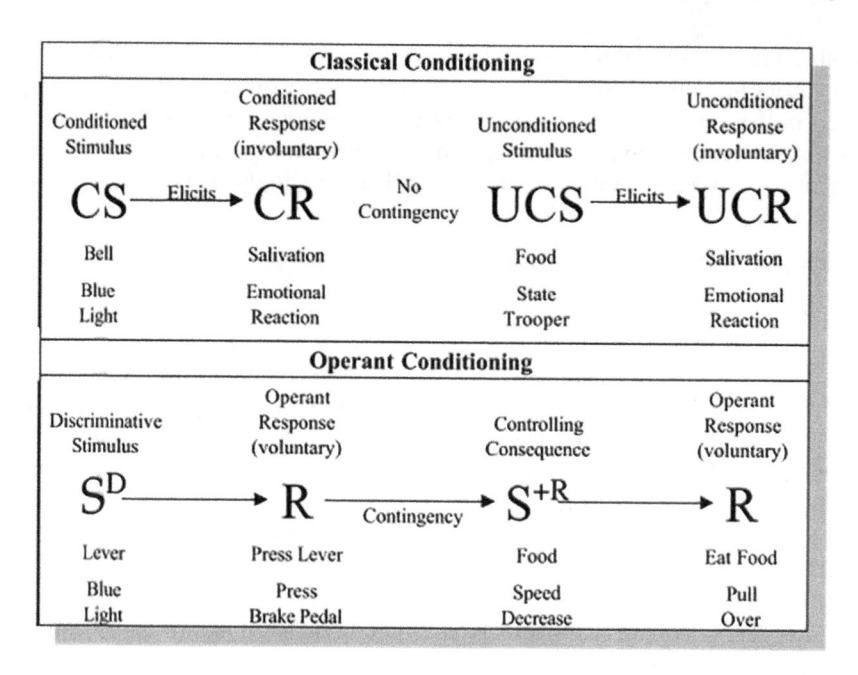

Operant Conditioning				
Discriminative Stimulus	Operant Response (voluntary)		Controlling Consequence	Operant Response (voluntary)
S^D —→	R	Contingency	S^{+R} —→	R
Lever	Press Lever		Food	Eat Food
Blue Light	Press Brake Pedal		Speed Decrease	Pull Over

FIGURE 7.4 The stimulus-response relationships in classical and operant conditioning overlap.

7.2.2 Operant Conditioning

As shown in the lower portion of Figure 7.4, the flashing blue light on the police car might influence you to press your brake pedal and pull over. That is not an automatic reflexive action but is a voluntary behavior you would perform in order to do what you consider appropriate at the time. In other words, you have learned (perhaps indirectly from watching or listening to others or from prior direct experience) to emit certain behaviors when you see a police car with a flashing blue light in your rearview mirror. Of course, you have also learned that certain driving behaviors (like pressing the brake pedal) will give a desired consequence (like slowing down the vehicle), and this will enable another behavior (like pulling over for an anticipated encounter with a police officer).

Selection by consequences. B. F. Skinner (1904–1990), the Harvard professor who pioneered the behavior-based approach to solving societal problems, studied this type of learning by systematically observing the behaviors of rats and pigeons in an experimental chamber referred to as a "Skinner Box" (much to Skinner's dismay). Dr. Skinner termed the learned behaviors in this situation "operant" because they were not involuntary and reflexive, as in classical conditioning; instead the subjects *operated* on the environment to obtain a certain consequence. A key principle demonstrated in the operant learning studies is that voluntary behavior is strengthened (increased) or weakened (decreased) by consequences (events immediately following behaviors).

The relation between a response and its consequence is a contingency, and this connection explains our motivation for doing most everything we choose to do. Thus, the hungry rat in the Skinner Box (or operant chamber) presses the lever to receive food, and the vehicle driver pushes the brake pedal to slow down the vehicle. Indeed, we all select various responses to perform daily—like eating, walking, reading, working, writing, talking, and recreating—to receive the immediate consequences they provide us. Sometimes, we emit the behavior to achieve a pleasant consequence—a reward. Other times, we perform a particular act in order to avoid an unpleasant consequence—a punitive consequence—and we usually stop performing behaviors that are consistently followed by a negative consequence.

Basic Principles   77

FIGURE 7.5 Activators direct behavior change when backed by a consequence.

The ABC (activator-behavior-consequence) contingency is illustrated in Figure 7.5. The dog will move if he expects to receive food after hearing the sound of the can opener. In other words, the direction provided by an activator is likely to be followed when it is backed by a consequence that is soon, certain, and significant. That is operant conditioning.

Might the dog in Figure 7.5 salivate when hearing the can opener? If the sound of the can opener elicits a salivation reflex in the dog, we have an example of classical conditioning. In this case, the can-opener sound is a CS and the salivation is the CR. What is the UCS in this example? Right on: the food which previously followed the sound of the electric can opener is the UCS, which causes the UCR of salivating without any learning experience. That UCS–UCR reflex is natural or "wired in" the organism.

An important point illustrated in Figure 7.5 is that operant and classical conditioning often occur simultaneously. While we operate on the environment to achieve a positive consequence or avoid a negative consequence, emotional reactions are often classically conditioned to specific stimulus events in the situation. We learn to like or dislike the environmental context and/or the people involved as a result of the type of ABC contingency influencing our ongoing behavior. That is how attitude can be negatively affected by enforcement techniques.

Emotional reactions. As indicated earlier in Chapter 3, we feel better when working for a pleasant consequence than when working to avoid or escape an unpleasant consequence. This can be explained by considering the classical conditioned emotions that naturally accompany the agents of positive vs. negative consequences. How do you feel, for example, when a police officer flashes a blue light to signal you to pull over? When the instructor asks to see you after class? When you enter the emergency area of a hospital? When the dental assistant motions that "you're next?" When your boss leaves you a phone message to "see him immediately?" When you see your family at the airport after returning home from a long trip?

Your reactions to those situations depend, of course, on prior experiences. Police officers, teachers, doctors, dentists, and supervisors do not typically elicit negative emotional reactions in young children. However, through the association of certain cues with the consequences we experience in situations involving these individuals, a negative emotional response or attitude can develop, and you do not have to experience these relationships directly. In fact, we undoubtedly learn more from observing and listening to others than from first-hand experience. This brings us to the third way in which we learn from experience—observational learning.

7.2.3 OBSERVATIONAL LEARNING

Substantial psychological science indicates that this form of learning is involved to some degree in almost everything we do. Whenever you do something a particular way because you saw someone else do it that way, because someone showed you to do it that way, or because characters on television or in a video game did it that way, you are experiencing observational learning. Whenever you attribute "peer pressure" as the cause of someone embracing an unhealthy habit (like smoking cigarettes or drinking excessive amounts of alcohol) or practicing an at-risk behavior (like driving at excessive speeds or adjusting equipment without locking out the energy source), you are referring to observational learning. When you remind someone to set an example for others, you are alluding to the decisive influence of observational learning.

Vicarious consequences. As children, we learned numerous behavioral patterns by watching our parents, teachers, and peers. When we saw our siblings or schoolmates receive rewards like special attention for certain behaviors, we were more likely to copy that behavior. This process is termed vicarious, or indirect, reinforcement. At the same time, when we observed others receiving a punitive consequence for emitting a certain behavior, we learned to avoid that behavior. This is referred to as vicarious punishment.

As adults, we teach others by example. As illustrated in Figure 7.6, our children learn new behavior patterns, including verbal behaviors, by watching us and listening to us. In this way they learn what is expected of them in various situations. I have talked with many parents of teenagers who are nervous about their son or daughter getting a driver's license. For some, their concern goes beyond the numerous dangers of real-world driving situations. They realize that several of their own driving behaviors, practiced regularly in front of their children for years, have not been exemplary.

FIGURE 7.6 Children learn a lot from their parents through observational learning.

Basic Principles

How can we expect our teenagers to practice safe driving and keep their emotions under control if we have shown them the opposite throughout their childhood? Of course, we are not the only role models who influence our children through observational learning, but we do make a difference.

Our actions influence others to a greater extent than we realize. Without being aware of our influence, children learn by watching us at home; our coworkers are influenced by our practices at work. Not only does the occurrence of safe acts encourage similar behavior by observers, but verbal behavior can also be influential. If a supervisor is observed commending a worker for her safe behavior or reprimanding an employee for an at-risk practice, observers may increase their performance of similar safety-related behaviors (through vicarious reinforcement) or decrease the frequency of similar at-risk behavior (through vicarious punishment). Indeed, to make safe behavior the norm—rather than the exception—we must always set an example both in our own work practices and in the verbal consequences we offer coworkers following their safe and at-risk behaviors. Figure 7.7 offers a memorable pictorial regarding the influence of example-setting on observational learning.

FIGURE 7.7 Intentionally and unintentionally, we teach through our example.

7.2.4 Overlapping Types of Learning

Laboratory methodologies have been able to study each type of learning separately, but the real world rarely offers such purity. In life, the usual situation includes simultaneous influence from more than one learning type. The operant learning situation, for example, is likely to include some classical (emotional) conditioning. As indicated earlier, this is one reason positive consequences should be used more frequently than negative consequences to motivate behavior change.

Remember, a rewarding situation (unconditioned stimulus) can elicit a positive emotional experience (unconditioned response), and a punitive situation (UCS) can elicit a negative emotional reaction (UCR). With sufficient pairing of positive or negative consequences with environmental cues (such as a work setting or particular people), the environmental setting (conditioned

FIGURE 7.8 Three types of learning occur in some situations.

stimulus) can elicit a positive or negative emotional reaction or attitude (conditioned response). This can in turn facilitate (if it is positive) or inhibit (if it is negative) ongoing performance.

Figure 7.8 depicts a situation in which all three learning types occur at the same time. As discussed earlier and diagrammed in Figure 7.4, the blue flashing light of the police car signals drivers to press the brake pedal of their car and pull over. In this case, the blue light is considered a *discriminative stimulus* because it tells people when to respond in order to receive or avoid a consequence. In this situation, the driver applied the brakes to avoid a punitive consequence, illustrating an avoidance contingency whereby the driver responded to avoid failure.

The flashing blue light might also serve as a conditioned stimulus, eliciting a negative emotional reaction. This is an example of classical conditioning occurring simultaneously with operant learning. A negative emotional reaction to the blue light might have been strengthened by prior observational learning. As a child, we might have seen one of our parents pulled over by a state trooper and subsequently observed a negative emotional reaction from our parent. The children in Figure 7.8 are not showing the same emotional reaction as the driver. Eventually, they will probably do so as a result of observational learning. Later, their direct experience as vehicle drivers will strengthen this negative emotional response to a flashing blue light on a police car.

7.3 IN SUMMARY

This chapter reviewed the basic principles underlying a behavior-based approach to the prevention and treatment of human problems. The variety of successful applications of this approach was discussed, based on personal experiences. Behavior-based principles—the primacy of behavior, direct assessment and evaluation, intervention by managers and peers, and three types of learning—were explained with particular reference to reducing personal injury.

Because at-risk behavior contributes to most injuries, a TSC requires a decrease in occurrences of at-risk behavior. Organizations have attempted to do this by targeting at-risk acts, and using corrective feedback, reprimands, or disciplinary action to motivate behavior change. That approach is useful but less proactive and less apt to be widely appreciated than a behavior-based approach that emphasizes recognition of safe behaviors. It would be easier to get employees involved in OSH processes if credits were given more often for doing the right thing than reprimands delivered for doing things wrong.

Three types of learning are relevant for understanding safety-related behaviors and attitudes. Most of our safe and at-risk behaviors are learned operant behaviors, performed in particular settings to gain positive consequences or to avoid negative consequences. Classical conditioning often occurs at the same time, linking a positive or negative emotional reaction with the stimulus cues surrounding the experience of receiving a positive or negative consequence. Those cues include the people who deliver the rewards or penalties.

We often learn what to do and what not to do by watching others receive recognition or correction for their operant behaviors. That is observational learning, an ongoing process that should inspire us to try to set the safe example at all times.

8 Identifying Critical Behaviors

The practical "how to" qualities of this book begin with this chapter. The overall process is called DO IT, each letter representing the four basic components of a behavior-based approach: Define target behaviors to influence, Observe those behaviors, Intervene to increase or decrease the occurrence of target behaviors, and Test the impact of your intervention process. This chapter focuses on developing a critical behavior checklist (CBC) for objective observing, intervening, and testing.

> As I grow older, I pay less attention to what men say, I just watch what they do.
>
> **(Andrew Carnegie)**

Now the action begins. Up to this point, I have been laying the groundwork—the rationale and theory—for the intervention strategies described here and in the next three chapters. From this information, you will learn how to develop action plans to increase occurrences of safe behaviors, decrease occurrences of at-risk behaviors, and cultivate a TSC.

Why did it take so long to get here—to the implementation stage? Indeed, if you are looking for "quick-fix" tools to make a difference in OSH, you may have skipped or skimmed the first two parts of this text and started your careful reading here. I appreciate that the pressures to get to the bottomline quickly are tremendous, but remember, there is no quick fix for OSH. The behavior-based approach that is the heart of this book is the most efficient and effective route to achieving a TSC. It is a never-ending continuous improvement process, one that requires ongoing and comprehensive involvement from the people protected by the process. In industry, these are the operators or line workers.

Long-term employee participation requires understanding and believing the principles behind the process. Employees must also perceive that they "own" the procedures that make the process work. For this to happen, it is necessary to teach the principles and rationale first (as organized in this book) and then work with participants to develop specific process procedures. This creates the perception of ownership and leads to long-term engagement.

When people are educated about the principles and rationale behind an OSH process, they can customize specific procedures for their particular work areas. Then the relevance of the training process becomes obvious, and participation is enhanced. People are more likely to accept and follow procedures they helped to create. They see such safe operating procedures as "the best way to do it" rather than "a policy we must obey because management says so."

As we begin here to define principles and guidelines for action plans, it is important to keep one thing in mind: You need to start with the conviction that there is rarely a generic best way to implement a process involving human interaction. For a behavior-based safety process to succeed at your setting, you will need to work out the procedural details with the people whose involvement is necessary. The BBS process needs to be customized to fit your culture.

8.1 THE DO IT PROCESS

For more than four decades, I have taught applications of the behavior-based approach to OSH with the acronym DO IT. The process is continuous and involves the following four steps.

Identifying Critical Behaviors

D: Define the critical target behavior(s) to increase or decrease in frequency of occurrence.
O: Observe the target behavior(s) during a pre-intervention baseline phase to set behavior-change goals and, perhaps, to understand natural environmental and/or social factors influencing the target behavior(s).
I: Intervene to change the target behavior(s) in desired directions of occurrence.
T: Test the impact of the intervention procedure by continuing to observe and record occurrences of the target behavior(s) during and after the Intervention phase.

From data obtained in the Test phase, you can evaluate the impact of your intervention and make an informed decision whether to continue it, implement another intervention strategy, or define another behavior to target for the DO IT process.

To begin, define clearly and concisely one or more target behaviors. This is the first step in the DO IT process. There is so much to choose from: using equipment safely, lifting correctly, locking out power appropriately, and looking out for the safety of others, to name just a few. The outcome of certain behaviors, such as wearing PPE, working in a clean and organized environment, and using a vehicle safety belt, could also be targeted.

If two or more people independently obtain the same frequency recordings when observing the defined target behavior or behavioral outcome during the same time period, you have a definition sufficient for an effective DO IT process. Baseline observations of the target behavior should be recorded before implementing an intervention program. More details on this aspect of the process are given later in this chapter.

What about the Intervention phase? This step of DO IT involves one or more behavior-change techniques, based on the simple ABC model depicted in Figure 8.1. As discussed in the preceding chapter, activators direct behavior and consequences motivate behavior. For example, a ringing telephone or doorbell activates the need for certain behaviors from residents, but residents answer or do not answer the telephone or door depending on current motives or expectations developed from prior experience.

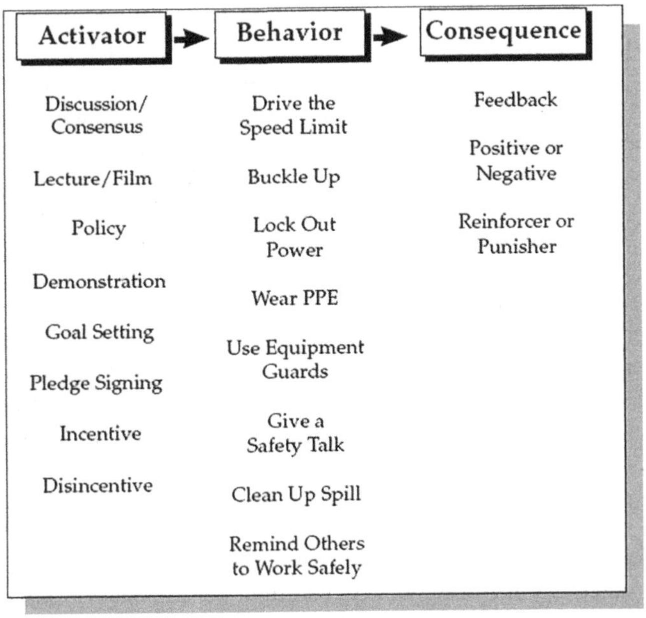

Activator	Behavior	Consequence
Discussion/Consensus	Drive the Speed Limit	Feedback
Lecture/Film	Buckle Up	Positive or Negative
Policy	Lock Out Power	Reinforcer or Punisher
Demonstration	Wear PPE	
Goal Setting	Use Equipment Guards	
Pledge Signing		
Incentive	Give a Safety Talk	
Disincentive	Clean Up Spill	
	Remind Others to Work Safely	

FIGURE 8.1 The ABC model is used to develop behavior change interventions.

FIGURE 8.2 Compared to at-risk behavior, safe behavior is often uncomfortable, inconvenient, and less fun.

Let's consider some motivating consequences related to OSH. The strongest consequences are soon, sizable, and certain. In other words, we work diligently for immediate, probable, and large positive consequences or rewards, and we work frantically to escape or avoid soon, certain, and sizable negative consequences or penalties. This helps explain why safety is a struggle in many workplaces. Safe behaviors are usually not supported by soon, certain, positive consequences. In fact, safe behaviors are often punished by soon, certain, negative consequences, including inconvenience, discomfort, and slower goal attainment. Also, the consequences that motivate safety professionals to promote safe work practices—reduced injuries and associated costs—are delayed, negative, and uncertain (actually improbable) from an individual's perspective.

Check out the two lawn-mower operators in Figure 8.2. Which one is having more fun? Who is more uncomfortable? Who is safe? Chances are both men will complete mowing their lawns without an injury. So which worker will have enjoyed the task more? This illustrates the fight against human nature discussed earlier in Part 2. Safety typically means more discomfort, inconvenience, and less fun than the more efficient at-risk alternative.

The DO IT process is a tool to use in this struggle with human nature. Developing and maintaining safe work practices often requires intervention strategies to keep people safe—interventions involving activators, consequences, or both. However, we are getting ahead of ourselves. First, we need to define critical behaviors to target for our intervention. Let's see how this is done.

8.2 DEFINING TARGET BEHAVIORS

The DO IT process begins by defining critical behaviors to address. These become the targets of our intervention strategies. Some target behaviors might be safe behaviors you want to see happen more often, like lifting with knees bent, cleaning a work area, putting on PPE, or replacing safety guards on machinery. Other target behaviors could be at-risk behaviors that need to be decreased in frequency, such as misusing a tool, overriding a safety switch, placing obstacles in an area designated for traffic flow, stacking materials incorrectly, and so on.

Identifying Critical Behaviors

A DO IT process can define desirable behaviors to be encouraged or undesirable behaviors to be changed. The focus of the process in your workplace depends on a review of your safety records, job hazard analyses, close-call reports, audit findings, interviews with employees, and other useful information.

Deciding which behaviors are critical is the first step of a DO IT process. A great deal can be discovered by examining the workplace and discussing with workers the various aspects of their jobs. Employees already know a lot about the hazards of their work and the safe behaviors needed to avoid an injury. They even know which safety policies are sometimes ignored to get the job done on time. They often know when a close call had occurred because an at-risk behavior or environmental hazard had been overlooked. They also know what at-risk behaviors could lead to a serious injury (or fatality) and what safe behaviors could prevent a serious injury (or fatality).

In addition to employee discussions, injury records and close-call reports can be consulted to discover critical behaviors (both safe and at-risk). Job hazard analyses or standard operating procedures can also provide information relevant to selecting critical behaviors to target in a DO IT process. Obviously, the plant safety director or the person responsible for maintaining records for OSHA or MSHA can provide valuable assistance in selecting critical behaviors to target.

After selecting target behaviors, it is critical to define them in a way that gets everyone on the same page. All participants in the process need to understand exactly what behaviors you intend to increase or decrease in frequency of occurrence. Defining target behaviors results in an objective standard for evaluating an intervention process.

8.2.1 What Is Behavior?

It is critical to define behaviors correctly. Let's begin by considering: What is behavior? Behavior refers to acts or actions by individuals that can be observed by others. In other words, behavior is what a person does or says as opposed to what s/he thinks, feels, or believes.

Yes, the act of saying words such as "I am tired," is a behavior because it can be observed or heard by others. However, this is not an observation of tired behavior. If the person's work activity slows down or the amount of time on the job decreases, we might infer that the person is actually tired. On the other hand, a behavioral "slow down" could result from other internal causes, like worker apathy or lack of interest. The important point here is that feelings, attitudes, or motives should not be confused with behavior. Those are *internal* dynamics of the person that cannot be directly observed by others. It is risky to infer internal person-states from external behavior.

8.2.2 Describing Behaviors

A target behavior needs to be defined in observable terms so multiple observers can independently watch one individual and obtain the same results regarding the occurrence or nonoccurrence of the target behavior. There should be no room for interpretation. "Is not paying attention," "acting careless," or "lifting safely," for example, are not adequate descriptions of behavior, because observers would not agree consistently about whether that behavior occurred. In contrast, descriptions like "keeping hand on handrail," "moving knife away from body when cutting," and "using knees while lifting" are objective and specific enough to obtain reliable information from trained observers. In other words, if two observers watched independently for the occurrence of those behaviors, they would likely agree on whether or not the target behavior occurred.

8.2.3 Multiple Behaviors

Let's examine types of behavior more closely. Some workplace activities can be treated effectively as a single behavior. Examples include: "Looking left–right–left before crossing the road," "Walking within the yellow safety lines," "Honking the fork-lift horn at the intersection," "Returning tools to their proper location," "Bending knees while lifting," and "Keeping a hand on the handrail while climbing stairs."

Some outcomes of behavior can also be targeted with definitions, like "using ear plugs," "using a vehicle safety belt," "climbing a ladder that is properly tied off," "working on a scaffold with appropriate fall protection," and "repairing equipment that had been locked out correctly." With a proper definition, an observer could readily count occurrences of those safe behaviors (or outcomes) during a systematic auditing process.

However, many safety activities include more than one discrete behavior, and it may be important to treat those behaviors independently in a definition for an audit. "Bending knees while lifting," for example, is only one component of a safe lift. Thus, if safe lifting were the activity targeted in a DO IT process, it would be necessary to define the separate behaviors (or procedural steps) of a safe lift. This would include, at least, checking the load before lifting, asking for help in certain situations, lifting with the legs, holding the load close to one's body, lifting in a smooth motion, and moving feet when rotating (or not twisting).

Each of the procedural steps in safe lifting requires a clear objective definition so two observers could determine independently and reliably whether the behaviors in a lifting sequence had occurred. Observing reliably whether the load was held close and knees were bent would be relatively easy. However, defining a "smooth lift" so that observers could agree on 80 percent or more of the observations would be more difficult. Plus, for observers to reliably audit "asking for help," the "particular situations" calling for such a request would need to be specified.

8.3 OBSERVING BEHAVIOR

The acronym "SOON" depicted in Figure 8.3 reviews the key aspects of developing adequate definitions of critical behaviors to target for a DO IT process. You are ready for the observation phase when you have a checklist of critical behaviors with definitions that are specific, observable, objective, and naturalistic. We have already considered most of the characteristics of behavioral definitions implied by these key words, and examples of behavioral checklists are provided later in this chapter, as well as in Chapter 12 on safety coaching.

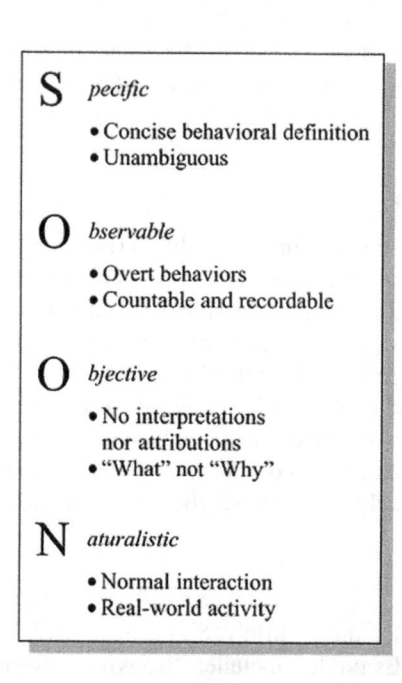

S *pecific*
- Concise behavioral definition
- Unambiguous

O *bservable*
- Overt behaviors
- Countable and recordable

O *bjective*
- No interpretations nor attributions
- "What" not "Why"

N *aturalistic*
- Normal interaction
- Real-world activity

FIGURE 8.3 Behavioral observations for the DO IT process should be "SOON."

Identifying Critical Behaviors

8.3.1 A Personal Example

Three decades ago, my daughter Krista asked me to drive her to the local Virginia Department of Motor Vehicles office to get her "learner's permit." She was 15 years old and thought she was ready to drive. Of course, I knew better, but how do you fight a culture that puts teenagers behind the wheel of motor vehicles before they are really ready for such a risky situation?

"Don't worry, Dad," my daughter said, "I've had driver's education in high school." Actually, that was part of my worry. She was educated about the concepts and rules regarding driving, but she had not been trained. She had not yet translated her education into operations or action plans.

In those days, to obtain a license to operate a motor vehicle in Virginia before the age of 18, teenagers with a learner's permit were required to take seven two-hour instructional periods of on-the-road experience with an approved driver-training school. For one-half of those sessions the student must be the driver; the rest of the time, the students they sit in the back seat and perhaps learn through observation. Thus, for seven one-hour sessions, Krista drove around town with an instructor in the front seat and one or more students in the back, waiting for their turn at the wheel. That was an opportunity for my daughter to transfer her driver education knowledge into actual performance.

Driving activators and consequences. On-the-job training obviously requires an appropriate mix of observation and feedback from an instructor. Practice does *not* make perfect. Only through appropriate feedback can people improve their performance. Some tasks give natural feedback to shape our behavior. When we turn a steering wheel in a particular direction, we see immediately the consequence of our action, and our steering behavior is naturally shaped. The same is true for several other behaviors involved in driving a motor vehicle—from turning on lights, windshield wipers, and cruise-control switches to pushing gas and brake pedals.

However, many other aspects of driving are not followed naturally with feedback consequences, particularly those that can prevent injury from vehicle crashes. Although we get feedback to tell us our steering wheel, gas pedal, turn signal lever, and brakes work, we do not get natural feedback regarding our safe vs. at-risk use of such control devices. Therefore, as shown in Figure 8.4, feedback must be added to the driving situation if we want behavior to improve.

FIGURE 8.4 Practice requires feedback to make perfect.

Also, when we first learn to drive, we do not readily recognize the activators that should signal the use of various vehicle controls. This is commonly referred to as "judgment." From a behavior-based perspective, "good driving judgment" is recognizing environmental conditions (or activators) that signal certain vehicle-control behaviors, and then implementing the controls properly.

I wondered whether my daughter's driving instructor would give her appropriate and systematic feedback regarding her driving "judgment." Would he point out consistently the activators that signal safe vehicle-control behaviors? Would he put emphasis on the positive by supporting my daughter's safe behaviors before correcting her at-risk behaviors? Would he, or a student in the back seat, display negative emotional reactions in certain situations and teach Krista (through classical conditioning) to feel anxious or fearful in particular driving scenarios? Would some at-risk driving behaviors by Krista or the other student drivers be overlooked by the instructor and lead to observational learning that some at-risk driving behaviors are acceptable?

Developing a Critical Behavior Checklist for driving. Even if the driving instruction is optimal, seven hours of such observation and feedback is certainly not sufficient to teach safe driving behaviors. I recognized a need for additional driving instruction for my daughter. We needed a DO IT process for driving. The first step was to define critical behaviors to target for observation and feedback. Through one-on-one discussion, my daughter and I derived a list of critical driving behaviors, and then we agreed on specific definitions for each item. My university students practiced using this critical behavior checklist a few times with various drivers and refined the list and definitions as a result. The CBC we eventually used is depicted in Figure 8.5.

Critical Behavior Checklist for Driving

Driver:	Date:	Day:
Observer 1:	Origin:	Start Time:
Observer 2:	Destination:	End Time:
Weather:		
Road Conditions:		

Behavior	Safe	At-Risk	Comments
Safety Belt Use:			
Turn Signal Use:			
Left turn			
Right turn			
Lane change			
Intersection Stop:			
Stop sign			
Red light			
Yellow light			
No activator			
Speed Limits:			
25 mph and under			
25 mph- 35 mph			
35 mph- 45 mph			
45 mph- 55 mph			
55 mph- 65 mph			
Passing:			
Lane Use:			
Following Distance (2 sec):			
Totals:			

% Safe = Total Safe Observations / (Total Safe + At-Risk Obs.) = _____ %

FIGURE 8.5 A critical behavior checklist can be used to increase safe driving.

Identifying Critical Behaviors

8.3.2 Using the Critical Behavior Checklist

After refining the CBC and discussing the final behavioral definitions with Krista, I felt ready to implement the second stage of DO IT—observation. I asked my daughter to drive me to Virginia Tech—about nine miles from home—to pick up some papers. I made it clear I would be using the CBC on both parts of the roundtrip. When we reached the university parking lot, I totaled the safe and at-risk checkmarks and calculated the percentage of safe behaviors. Krista was quite anxious to learn the results, and I looked forward to giving her objective behavioral feedback. I had good news. Her percentage of safe driving behaviors (percent safe) was 85 percent, and I considered that quite good for our first time.

I told Krista her "percent safe" score and proceeded to show her the list of safe checkmarks while covering the checks in the At-Risk column. Obviously, I wanted to make this a positive experience, and to do this, it was necessary to emphasize the behaviors I saw her perform correctly. To my surprise, she did not seem impressed with her 85 percent safe score, and she pushed me to tell her what she had done wrong. "Get to the bottomline, Dad," she asserted, "Where did I screw up?" I continued an attempt to make the experience positive by saying, "You did great, Honey, look at the high number of safe behaviors." "But why wasn't my score 100 percent?" reacted Krista. "Where did I go wrong?"

This initial experience with the driving CBC was enlightening in two respects. It illustrated the unfortunate reality that the "bottomline" for many people is: "Where did I make a mistake?" My daughter, at age 15, had already learned that people evaluating her performance seem to be more interested in mistakes than successes. That obviously makes performance evaluation (or appraisal) an unpleasant experience for many people.

A second important outcome from this initial CBC experience was the realization that people can be unaware of their at-risk behaviors, and only through objective behavioral feedback can that be changed. My daughter did not readily accept my corrective feedback regarding her four at-risk behaviors. In fact, she vehemently denied that she did not always come to a complete stop. However, she was soon convinced of her error when I showed her my data sheet and my comment regarding the particular intersection where there was no traffic and she made only a rolling stop before turning right. I did remind her that she did use her turn signal at that and every intersection, and this was something to be proud of. She was developing an important safety routine, one often neglected by many drivers.

Frankly, I did not appreciate those two lessons from that first application of the driving CBC until my daughter monitored my driving. Yes, Krista used the CBC in Figure 8.5 to evaluate my driving on several occasions. I found this reciprocal application of a CBC to be most useful in developing mutual trust and understanding between us. I found myself asking my daughter to explain my lower-than-perfect score and arguing about one of the recorded "at-risk" behaviors. I, too, was defensive about not being 100 percent safe. After all, I had been driving for 37 years and teaching and researching safety for more than 20 years. How could I not get a perfect driving score when I knew I was being observed?

From our experience with the CBC, my daughter and I learned the true value of an observation-and-feedback process. While using the checklist does transfer education into training through systematic observation and feedback, the real value of the process is the interpersonal coaching (i.e., the behavioral feedback) that occurs. In other words, we learned not to get too hung up on the actual numbers. After all, there is plenty of room for error in the numerical scores. Rather, we learned to appreciate the fact that through this process people are actively caring for the safety and health of each other in a way that can truly make a difference. We also learned that even experienced participants can perform at-risk behavior and not even realize it.

8.4 TWO BASIC APPROACHES

The CBC examples described previously illustrate two basic approaches to implementing the Define and Observe stages of DO IT. The driving CBC I developed with my daughter illustrates the observation-and-feedback process recommended by a number of successful BBS consultants. I refer to this approach as one-to-one safety coaching because it involves an observer using a CBC to provide instructive behavioral feedback for another person.

The second approach to the Define and Observe stages of DO IT involves a limited CBC (perhaps targeting only one behavior) and does not necessarily involve one-to-one coaching. This is the approach used in most of the published studies of the behavior-based approach to OSH.

Each of those approaches to the Define and Observe stages of DO IT are advantageous for different applications within the same culture. Thus, it is important to understand the basic procedures of each and to consider their advantages and disadvantages. For some work settings, I have found it quite useful to start with the simpler approach of targeting only a few CBC behaviors. With immediate success, behaviors are then added until eventually a comprehensive CBC is developed, accepted, and used willingly throughout a worksite.

8.4.1 STARTING SMALL

This approach targets a limited number of critical behaviors but does not require one-on-one observation. A work group defines a critical behavior or behavioral outcome to observe, as discussed earlier in this chapter. After defining their target behaviors so that two or more observers can reliably observe and record a particular characteristic of the behavior, usually frequency of occurrence, the group members should give each other permission to observe this work practice among themselves. If some group members do not give permission, it is best not to argue with them. Simply exclude those individuals from the observations and invite them to join the process whenever they feel ready. They will likely participate eventually when they see that the DO IT process is not a "gotcha program" but an objective and effective way to actively care for the safety of others and to cultivate a TSC.

It often helps to develop a CBC for use during observations. As discussed earlier, target behaviors like "safe lifting" and "safe use of stairs" should include a few specific behaviors, either safe or at-risk. Therefore, the CBC should list each behavior separately and include columns for checking "safe" or "at-risk." Figure 8.6 depicts a sample CBC for safe lifting. Through use of

Observer:	Date:	
Target Behavior	Safe	At-Risk
load appropriate		
hold close		
use legs		
move feet - don't twist		
smooth motion - no jerks		
Comments (use back if necessary):		

% Safe Observations:

$$\frac{\text{Total Safe Observations}}{\text{Total Safe Observations + At-Risk Observations}} \times 100 = \_\_\_\_\%$$

FIGURE 8.6 A critical behavior checklist can be used to increase safe lifting.

this CBC, a work group might revise the definitions and possibly add a lifting-related behavior relevant to their particular work area.

Participants willing to be observed anonymously for the target behavior(s) use the CBC to maintain daily records of the safe and at-risk behaviors defined by the group. They do not approach another individual specifically to observe him or her. Rather they look for opportunities for the target behavior to occur. When they see a safe behavior opportunity (SBO), they take out their checklist and complete it. If the target behavior is "safe lifting," for example, observers keep on the lookout for an SBO for lifting. They might observe such an SBO from their work station or while walking through the plant. Of course, if they see an at-risk lifting behavior and are close enough to reduce the risk, they should put their CBC aside and intervene. Intervening to reduce risk should take precedence over recording an observation of at-risk behavior.

8.4.2 Observing Multiple Behaviors

As the list of target behaviors on a CBC increases, it becomes more and more difficult to complete a checklist from a remote location. Auditing several critical behaviors usually puts observers in close contact with another person (the performer), resulting in a one-on-one coaching situation. The observer should seek permission from the performer before recording any observations, even though a work group might have agreed on the observation process in earlier education-and-training meetings. If the performer wishes not to be observed, the observer should leave with no argument and a friendly smile. This helps to build the trust needed to eventually reach 100 percent participation in the DO IT process.

Multiple-behavior CBCs might be specific to a particular job or be generic in nature. The driving CBC I used with Krista was a job-specific checklist, only relevant for operating a motor vehicle. In contrast, a generic checklist is used to observe behaviors that may occur at various job sites. The CBC depicted in Figure 8.7 is generic because it is applicable for any job that requires

Critical Behavior Checklist for Personal Protective Equipment

Observation period (dates): __________

Observer: __________

TOTAL NUMBER OF EMPLOYEES OBSERVED	NUMBER OF EMPLOYEES OBSERVED USING ALL REQUIRED PPE

PPE (For Observed Area)	SAFE OBSERVATION (Proper Use of PPE)	AT-RISK OBSERVATION (Improper or No Use of PPE)
Gloves		
Safety Glasses/Shield		
Hearing Protection		
Safety Shoes		
Hard Hat		
Lifting Belt		
TOTAL		

FIGURE 8.7 A critical behavior checklist can be used to increase the use of personal protective equipment.

the use of PPE. Because different PPE might be required on different jobs, certain PPE categories on the CBC may be irrelevant for some observations. For jobs requiring extra PPE, additional behaviors will be targeted on the CBC. Obviously, the observer needs to know the PPE requirements before attempting to use a CBC like the one shown in Figure 8.7.

The CBC in Figure 8.7 includes a place for the observer's name, but the performer's name is not recorded. Also, this CBC was designed to conduct several one-on-one behavioral audits over a period of time. Each time the observer performs an observation, s/he places a checkmark in the left box (for total number of observations). If the performer was using all PPE required in the work area, a check would be placed in the right-hand box. From these entries, the overall percentage of safe employees can be monitored.

The checkmarks in the individual behavior categories of the CBC in Figure 8.7 are totaled, and by dividing the total number of safe checks by the total safe and at-risk checks, the percentage of safe behaviors for each PPE category can be assessed. See the formula at the bottom of the CBC in Figures 8.5 and 8.6. This enables valuable feedback regarding the relative use of various devices to protect employees. Such information might suggest a need to make certain PPE more comfortable or convenient to use. It might also suggest the need for a special behavior-change intervention, as discussed in the next three chapters.

The formula at the bottom of the CBC in Figures 8.5 and 8.6 can be used to calculate an overall percent safe score. We have found it very effective to post this global score weekly for different work teams. Such social-comparison information presumably motivates performance improvement through friendly intergroup competition. Chapter 12 also includes additional information on the design of CBCs for one-on-one behavioral observations.

8.5 IN SUMMARY

This chapter explored the "nuts and bolts" of implementing a BBS process to achieve a TSC. The overall process is referred to as DO IT, each letter representing one of the four stages of BBS. This chapter focused on the first two stages—Define and Observe.

Defining critical behaviors to target for observation and intervention is not easy. A work team needs to consult a variety of sources, including the workers themselves, close-call reports, injury records, job-hazard analyses, and the plant safety director. After selecting a list of behaviors critical to preventing injuries in their work area, the team needs to struggle through defining those behaviors so precisely that all observers agree on a particular characteristic of each behavior at least 80 percent of the time. The behavioral property most often observed for OSH is frequency of occurrence per individual worker or per group of employees.

A critical behavior checklist is used to observe and record the relative frequency (or percentage of opportunities) that critical behaviors occur throughout a work setting. If the CBC contains only a few behaviors or behavioral outcomes (conditions caused by behavior), it is possible to conduct observations without engaging in a one-on-one coaching session. This is often the best approach to use when first introducing BBS to a work culture. It is not nearly as overwhelming or time-consuming as one-on-one coaching with a comprehensive CBC.

Over time and through building trust, a short CBC can be readily expanded and lead to one-on-one safety coaching. Safety coaching is one very effective way to implement each stage of the DO IT process and is detailed in Chapter 12. First, it is important to understand how the first two stages of DO IT can facilitate a proper behavioral analysis of the situation. This is the theme of the next chapter.

9 Behavioral Safety Analysis

The defining and observing processes of DO IT provide opportunities to evaluate the situational factors contributing to at-risk behavior and a possible injury. This chapter details the procedures of a behavioral safety analysis, including a step-by-step examination of the situational, social, and personal factors influencing at-risk behavior in order to determine the most cost-effective corrective action. Critical distinctions are made between four types of intervention—1) instruction, 2) motivation, 3) support, and 4) self-management—as well as between training and education, and between accountability and responsibility.

> A prescription without diagnosis is malpractice.
>
> (Socrates)

Chapter 8 introduced the DO IT process and provided some details about the first two steps—*define* target behavior(s) to improve and *observe* the target behavior occurring naturally in the work environment. The CBC was introduced as a way to look for occurrences of critical behaviors during a work routine and then offer workers one-on-one feedback about what behavior was safe and what behavior was at-risk. This is behavior-based coaching and is explained in more detail in Chapter 12.

While an observer is checking for safe and at-risk occurrences of critical behaviors, s/he is looking for contributing factors to at-risk behavior, as well as for the occurrence of behaviors that ought to be included in a revised CBC. The information in the "Comments" section of a CBC is invaluable for determining what factors contribute to at-risk behavior and should be changed to reduce occurrences of such behavior. Before intervening to correct a problem, a proper behavioral analysis is needed. That is the theme of this chapter. Without a careful behavioral analysis of the situation requiring an intervention or corrective action, accusations of malpractice could be warranted.

9.1 REDUCING BEHAVIORAL DISCREPANCY

It is important to consider human performance problems a discrepancy rather than a deficiency. This places the focus on the behavior, not the individual. In other words, a difference exists between the behavior demonstrated and the behavior desired. When evaluating OSH issues, this discrepancy is between behavior considered at-risk vs. safe.

The behavioral discrepancy could be an "error of omission" or an "error of substitution." The worker might have failed to perform a particular safe behavior because s/he took a shortcut, or the individual could have performed a certain behavior that puts someone at risk for an injury. After deciding what is safe and what is at-risk for a particular individual and work situation, an action plan can be designed to reduce the discrepancy between what is and what should be. Let's consider the variety of situations or work contexts that can influence a behavioral discrepancy.

9.1.1 Can the Task Be Simplified?

Before designing an intervention to reduce a behavioral discrepancy, make sure all possible engineering "fixes" have been implemented. For example, consider the many ways the environment could be changed to reduce physical effort, reach, and repetition. In other words, entertain ways to make the job more user-friendly before deciding what behaviors are needed to prevent an

FIGURE 9.1 We are reluctant to accept personal responsibility for our injuries.

injury. This is, of course, the rationale behind ergonomics and the search for engineering solutions for OSH.

As discussed earlier in Chapter 6, when people experience failure, as reflected by noncompliance, property damage, or personal injury, they are more likely to place blame on external than personal factors. In other words, as illustrated in Figure 9.1, people involved in an injury feel more comfortable discussing environment-related causes than person factors. Given this self-serving bias it makes sense to begin a behavioral analysis with a discussion of environmental or engineering factors. Afterwards, the possibility of a constructive discussion of human factors potentially contributing to the incident increases markedly.

Sometimes behavior facilitators can be added, such as:

1. Control designs with different shapes so they can be discriminated by touch as well as sight.
2. Clear instructions placed at the point of application.
3. Color codes to aid memory and task differentiation.
4. Convenient machine lifts or conveyor rollers to help with physical jobs.

Plus, complex assignments can sometimes be redesigned to involve fewer steps or more participants. To reduce boredom or repetition, simple tasks might allow for job swapping.

Ask these questions at the start of a behavioral analysis:

- Can an engineering intervention make the job more user friendly?
- Can the task be redesigned to reduce physical demands?
- Can a behavior facilitator be added to improve response differentiation, reduce memory load, or increase reliability?
- Can the challenges of a complex task be shared?
- Can boring, repetitive jobs be swapped?

Behavioral Safety Analysis

9.1.2 Is a Quick Fix Available?

Behavior might be more at-risk than desired because expectations are unclear, resources are inadequate, or behavioral feedback is unavailable. In these cases, solutions to reducing a behavioral discrepancy are obvious and relatively inexpensive. Behavior-based instruction or demonstration can overcome invisible expectations, and behavior-based feedback can enable continuous improvement. Furthermore, a work team could decide what resources are needed to make a safe behavior more convenient, comfortable, or efficient.

When conducting this aspect of a behavioral analysis, ask these questions:

- Does the individual know what safety precautions are expected?
- Are there obvious barriers to safe work practices?
- Is the equipment as safe as possible under the circumstances?
- Is PPI readily available and as comfortable as possible?
- Do employees receive behavior-based feedback related to OSH?

9.1.3 Is Safe Behavior Punished?

As explained in Chapter 8, a key principle of applied behavioral science is that behavior is motivated by its consequences. In other words, our behavior results in favorable or unfavorable consequences, and those consequences determine our future behavior. Sometimes naturally occurring consequences work against us. This is especially true for OSH because safe behavior is usually less comfortable, convenient, or efficient than the at-risk alternative.

Those analyzing an incident need to try to see the situation through the eyes of the performer. This is called empathy and is illustrated in Figure 9.2. Some consequences might actually seem positive to an observer but be viewed as negative by the performer. For example, a safety manager might consider an individual's public safety award a positive consequence, but for the individual it could be a negative consequence because of expected harassment from coworkers.

FIGURE 9.2 It is useful to see a situation through the eyes of the other person.

Public praise from supervisors and teachers can be overshadowed by negative consequences from coworkers and classmates, resulting in a reduction in individual effort.

In some work cultures, the interpersonal consequences for reporting an environmental hazard or close call are more negative than positive. After all, those situations imply that someone was irresponsible or careless. It is not unusual for people to be ridiculed for wearing protective gear or using an equipment guard. It might even be considered "cool" or "macho" to work unprotected and take risky shortcuts. The hidden agenda might be that "only a 'gutless chicken' would wear that fall protection."

Mager and Pipe (1997) refer to those situations as "upside-down consequences" and suggest that whenever a behavioral discrepancy exists, part of the problem is because the desired behavior is punished. Are people put down when they should be lifted up? Are the consequences for performing well perceived as negative rather than positive, as illustrated in Figure 9.3? It is not unusual for the best performers to be "rewarded" with extra work.

FIGURE 9.3 Sometimes exemplary performance is punished.

Ask these questions during your behavioral analysis:

- What are the consequences for safe behavior?
- What are the consequences for at-risk behavior?
- Are there more negative than positive consequences for safe behavior?
- What negative consequences for safe behavior can be reduced or removed?

9.1.4 Is At-Risk Behavior Rewarded?

As indicated previously, at-risk behavior is often followed by natural positive consequences. Shortcuts are usually taken to save time and can lead to a faster rate of output. So, taking on at-risk shortcuts can be labeled "efficient" behavior. I have analyzed several work environments, for example, where bypassing or overriding the power lockout switches was acceptable because it benefited production—the bottomline. In those cultures, the worker who could fix or adjust

Behavioral Safety Analysis

equipment without locking out the power was a hero. He could handle equipment problems without slowing down production.

Behavior does not occur in a vacuum. Most people perform the way they do because they expect to achieve soon, certain, and positive consequences or they expect to avoid soon, certain, and negative consequences. People take calculated risks because they expect to gain something positive or avoid something negative.

Ask these questions:

- What are the soon, certain, and positive consequences for at-risk behavior?
- Does a worker receive more attention, prestige, or status from coworkers for at-risk than safe behavior?
- What positive consequences for at-risk behavior can be reduced or removed?

9.1.5 Are Extra Consequences Used Effectively?

Because the natural consequences of comfort, convenience, and efficiency usually support at-risk over safe behavior, it is often necessary to add extra consequences. These usually take the form of incentive/reward or disincentive/penalty programs. Unfortunately, many of those programs do more harm than good because they are implemented ineffectively. Disincentives are often ineffective because they are used inconsistently and motivate avoidance behavior rather than achievement. Moreover, safety incentive programs based on *outcomes* stifle the development and administration of an effective safety incentive program to improve *behavior*. Details about designing an effective safety incentive/reward program are provided in Chapter 11.

Ask these questions when analyzing the impact of extra consequences put in place to motivate improved safety performance:

- Can the penalties for undesirable behavior be implemented consistently and fairly?
- Can the safety incentives stifle the reporting of injuries and close calls?
- Do the safety incentives motivate the achievement of OSH-process goals?
- Do monetary rewards foster participation only for a financial payoff and conceal the real benefit of safety-related behavior—injury prevention?
- Are workers recognized individually and/or as teams for completing process activities related to safety improvement?

9.1.6 Is There a Skill Discrepancy?

What about those times when the individual does not know how to do the prescribed safe behavior? The person is "unconsciously incompetent." This situation might call for training which is a relatively expensive approach to corrective action. Most of the time a behavioral discrepancy is not caused by a genuine lack of skill. Usually people can perform the safe behavior if the conditions and the consequences are right. So, training should really be the least-used approach for corrective action.

Ask these questions to determine whether the behavioral discrepancy is caused by a lack of skill:

- Could the person perform the task safely if his or her life depended on it?
- Are the person's current skills adequate for the task?
- Did the person ever know how to perform the job safely?
- Has the person forgotten the safest way to perform the task?

9.1.7 What Kind of Training Is Needed?

Answers to the last two questions can help pinpoint the kind of intervention needed to reduce a skill discrepancy. More specifically, a "yes" answer to those questions implies the need for a skill-maintenance program. Skill maintenance might be needed to help a person stay skilled, such as police officers practicing regularly on a pistol range to stay ready to use their guns effectively in the rare situation when they need them. This is, of course, the rationale behind periodic emergency training. People need to practice the behaviors that could prevent an injury or save a life during an emergency. Fortunately, emergencies do not happen very often, but since they do not, people need to go through the motions just to "stay ready to perform." Then, if the infrequent event does occur, they will be ready to do the right thing.

A very different kind of situation also calls for skill-maintenance training. This is when certain behaviors occur regularly, but discrepancies still exist. Contrary to circumstances requiring emergency training, this problem is not lack of practice. Rather, the person gets plenty of practice doing the behavior ineffectively or unsafely. In this case, practice does not make perfect but rather serves to entrench a bad (or at-risk) habit.

Vehicle driving behavior is perhaps the most common and relevant example of this second kind of situation in need of skill-maintenance training. Most drivers know how to drive a vehicle safely, and at one time they performed very few at-risk driving behaviors. However, for many drivers, safe driving has declined, with some safe-driving practices dropping out of some people's driving repertoire completely.

Practice with appropriate behavior-based feedback is critical for solving both types of skill discrepancies. However, if the skill is already used frequently but has deteriorated (as in the driving example), it is often necessary to add an extra feedback intervention to overpower the natural consequences that have caused the behavior to drift from the ideal.

Ask these questions to determine whether the cause of the apparent skill discrepancy is due to lack of practice or lack of feedback:

- How often is the desired skill performed?
- Does the performer receive regular behavior-based feedback relevant to skill maintenance?
- How does the performer find out how well s/he is doing?

9.1.8 Is the Person Right for the Job?

From this discussion, it is clear a skill discrepancy can be handled in one of two ways: change the job or change the behavior. The first approach is exemplified by simplifying the task, while the latter approach requires practice and behavior-based feedback or behavioral coaching. But what if a person's interests, skills, or prior experiences are incompatible with the job?

Ask these questions to determine whether the individual has the potential to handle the job safely and effectively:

- Does the person have the physical capability to perform as desired?
- Does the person have the mental capability to handle the complexities of the task?
- Is the person overqualified for the job and thus prone to boredom or dissatisfaction?
- Can the person learn how to do the job as desired?

9.1.9 In Summary

Figure 9.4 summarizes the main steps of a behavior-based incident analysis with a flow chart of ten basic questions to ask. Before an individual worker is targeted with a training intervention, engineering strategies are considered for task simplification.

Behavioral Safety Analysis

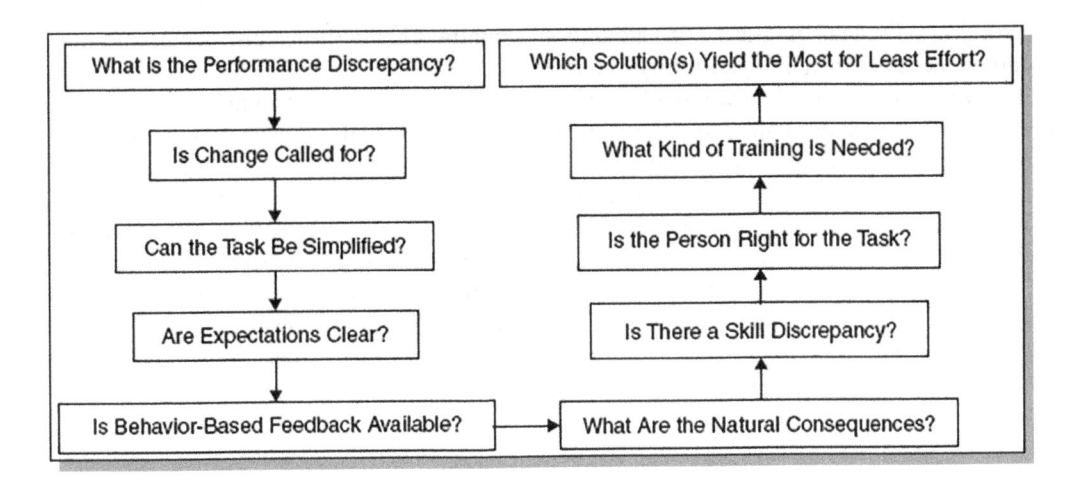

```
┌─────────────────────────────────────┬──────────────────────────────────────────┐
│  What is the Performance Discrepancy?│  Which Solution(s) Yield the Most for Least Effort?│
│               ↓                      │               ↑                          │
│      Is Change Called for?           │      What Kind of Training Is Needed?    │
│               ↓                      │               ↑                          │
│    Can the Task Be Simplified?       │     Is the Person Right for the Task?    │
│               ↓                      │               ↑                          │
│      Are Expectations Clear?         │      Is There a Skill Discrepancy?       │
│               ↓                      │               ↑                          │
│  Is Behavior-Based Feedback Available? → What Are the Natural Consequences?    │
└─────────────────────────────────────┴──────────────────────────────────────────┘
```

FIGURE 9.4 Ask ten basic questions to conduct a behavior-based incident analysis.

Bottomline: Before deciding on an intervention approach, conduct a careful analysis of the situation, the behavior, and the individual(s) involved in an observed discrepancy between the behavior desired and the behavior observed. Do not impulsively assume corrective action to improve behavior requires training or "discipline." A behavioral incident analysis will likely give priority to a number of alternative intervention approaches. Critical disadvantages of using "discipline" or a punitive approach to corrective action are explicated in Chapter 11.

9.2 BEHAVIOR-BASED SAFETY TRAINING

The principles and procedures of behavior-based safety, including behavioral observation and interpersonal coaching, are new to many people. Therefore, to achieve a TSC, BBS education and training are needed throughout a work culture. Everyone in a workforce needs to understand the basic rationale or theory behind the BBS approach. Then, work teams need to participate in exercises to customize observation, analysis, and feedback procedures for their work areas. Finally, practice sessions are needed in which individuals and teams receive supportive and corrective feedback regarding their implementation of BBS—from designing a CBC and analyzing CBC results to using a CBC for constructive intervention, especially for delivering supportive and corrective feedback.

Why should employees want such training? First, as indicated in earlier chapters, BBS works to reduce injuries. The principles and methods of BBS are applicable in many situations—whenever and wherever human behavior is a factor and can be improved. Thus, training in BBS provides skills useful in numerous situations at work, at home, during recreational and sport activities, and traveling in between.

While people need sufficient training to feel confident they can complete a certain task they also need to believe the job is worthwhile. This requires education, not training. There is a difference. Actually, you already know the distinction. Do you want your teenager to receive sex education or sex training? In contrast, are you satisfied if your teenager receives only "driver education," or do you prefer some "training" with that education?

Because people know intuitively the difference between education and training, misusing these terms can lead to problems. Safety training might be perceived as a step-by-step procedure or program with no room for individual creativity, ownership, or empowerment.

This is how safety can come to be viewed as a top-down "flavor of the month." If we do not educate people about the principles or rationale behind a particular OSH policy, program, or process, individuals might participate only minimally. They will perceive the program as a requirement rather than an opportunity to make a difference.

Likewise, safety education without follow-up training will not reap optimal benefits. Learning the theory or principles behind an intervention approach is crucial for customizing intervention procedures for a particular work situation, but after the procedures are developed—with input from an educated work team—training is necessary. People need to know precisely what to do. With proper education, these participants can refine or upgrade procedures when appropriate; and following a change in procedures, additional training is obviously needed.

Bottomline: People need both education and training to improve. As Deming is known for reiterating at his quality and productivity workshops, "There's no substitute for knowledge" (1991, 1992). Indeed, without gaining profound knowledge through education and training, we are like the clown in Figure 9.5. We do our best with what we now know. We use our biased and nonoptimal commonsense.

FIGURE 9.5 Without education and training, we clown around with our biased commonsense.

9.3 INTERVENTION AND THE FLOW OF BEHAVIOR CHANGE

Taken together, education and training reflect instruction, and instruction represents a type of intervention. Under certain circumstances, instruction is sufficient to change behavior. Sometimes instruction does not work, and another type of intervention is needed. Perhaps a motivational intervention is called for, or perhaps only a supportive intervention is needed. A complete behavioral safety analysis should often include a recommendation for a certain type of behavior-focused intervention. This section provides information critical for making such a recommendation. Then, subsequent chapters provide guidance for designing a particular type of constructive intervention.

9.3.1 Three Types of Behavior

On-the-job behavior starts out as other-directed behavior, given participants follow someone else's instructions. Such direction can come from a training program, an operation's manual, or a policy statement. After learning what to do, essentially by memorizing or internalizing the appropriate instructions, behavior can become self-directed. Participants can talk to themselves or formulate an image before performing a behavior in order to activate the right response.

After performing some behaviors frequently and consistently over a period of time they become automatic or routine. A habit is formed. Some habits are good and some are not good, depending on their short and long-term consequences. If implemented correctly, rewards, recognition, and other positive consequences can facilitate the transfer of behavior from the other-directed phase to the self-directed phase.

Before a bad habit can be changed to a good habit, the target behavior must become self-directed. In other words, people need to become aware of their undesirable habit (as in at-risk behavior) before adjustment is possible. Then, if people are motivated to improve (perhaps as a result of an incentive/reward program), their new self-directed behavior can become automatic or routine.

9.3.2 Three Kinds of Intervention Strategies

Chapter 8 presented the ABC model as a framework to understand and analyze behavior, as well as to develop interventions for improving behavior. Recall that the "A" stands for activators or antecedent events that precede behavior or "B," and "C" refers to the consequences following "B"—the behavior. Recall that activators direct behavior, and consequences motivate behavior.

Instructional intervention. An instructional intervention is typically an activator or an antecedent event used to get new behavior started or to move behavior from the automatic (habit) stage to the self-directed stage. Or it is used to improve behavior already in the self-directed stage.

This type of intervention consists primarily of activators, as exemplified by education sessions, training exercises, and directive feedback or feedforward (cf. Geller & Geller, 2023). Because the purpose is to instruct, the intervention comes before the target behavior and focuses on helping the performer internalize the instructions. As we have all experienced, this type of intervention is more effective when the instructions are specific and given one on one. Role-playing exercises enable instructors to customize directions specific to an individual's attempts to improve. They also provide participants an opportunity to receive supportive feedback for their improvement.

Supportive intervention. Once a person learns the right way to do something, practice is important so the behavior becomes part of a natural routine. Continued practice leads to fluency and, in many cases, to automatic or habitual behavior. This is an especially desirable state for safety-related behavior, but practice does not come easily and benefits greatly from supportive intervention. We need support to reassure us we are doing the right thing and to encourage us to keep going.

While instructional intervention consists primarily of activators, supportive intervention focuses on the application of positive consequences. Thus, when we give people supportive feedback, recognition, or gratitude for a particular safe behavior, we are showing our appreciation for their efforts, and increasing the likelihood they will perform that behavior again. Each occurrence of the desired behavior facilitates fluency and helps to build a good habit.

Motivational intervention. When people know what to do and do not do it, a motivational intervention is needed. In other words, they require some external encouragement or pressure to change. Instruction alone is obviously insufficient because individuals are knowingly doing the wrong thing. As discussed earlier in Chapter 4, we refer to that behavior as taking a calculated risk.

We usually perform calculated risks because we perceive the positive consequences of the at-risk behavior to be more powerful than the negative consequences. This is because the positive consequences of comfort, convenience, and efficiency are immediate and certain, while the negative consequence of at-risk behavior (such as an injury) is improbable and seems remote. Furthermore, the safe alternative is relatively inconvenient, uncomfortable, or inefficient, and such negative consequences are immediate and certain.

Under these circumstances an incentive/reward intervention can be useful. Such a program attempts to motivate a certain target behavior by promising participants a positive consequence if they perform it. The promise is the incentive and the consequence is the reward. In OSH, this kind of motivational intervention is much less common than a disincentive–penalty contingency. This is when a rule, policy, or law threatens to give people a negative consequence (a penalty) if they fail to comply or take a calculated risk.

9.3.3 THE FLOW OF BEHAVIOR CHANGE

Figure 9.6 reviews this intervention information by depicting relationships among four competency states (unconscious incompetence, conscious incompetence, conscious competence, and unconscious competence) and four intervention approaches (instructional intervention, motivational intervention, supportive intervention, and self-management). When people are unaware of the safe work practice (i.e., they are "unconsciously incompetent"), they need repeated instructional intervention until they understand what to do. Then, as depicted at the far left of Figure 9.6, the critical question is whether they perform the desired behavior. If they do, the question of behavioral fluency is relevant. A fluent response becomes a habit or part of a regular routine, and thus the individual is "unconsciously competent."

When workers know how to perform a task safely but do not, they are considered "consciously incompetent" or irresponsible. This is when an external motivational intervention can be useful, as discussed previously. Then when the desired behavior occurs at least once, supportive

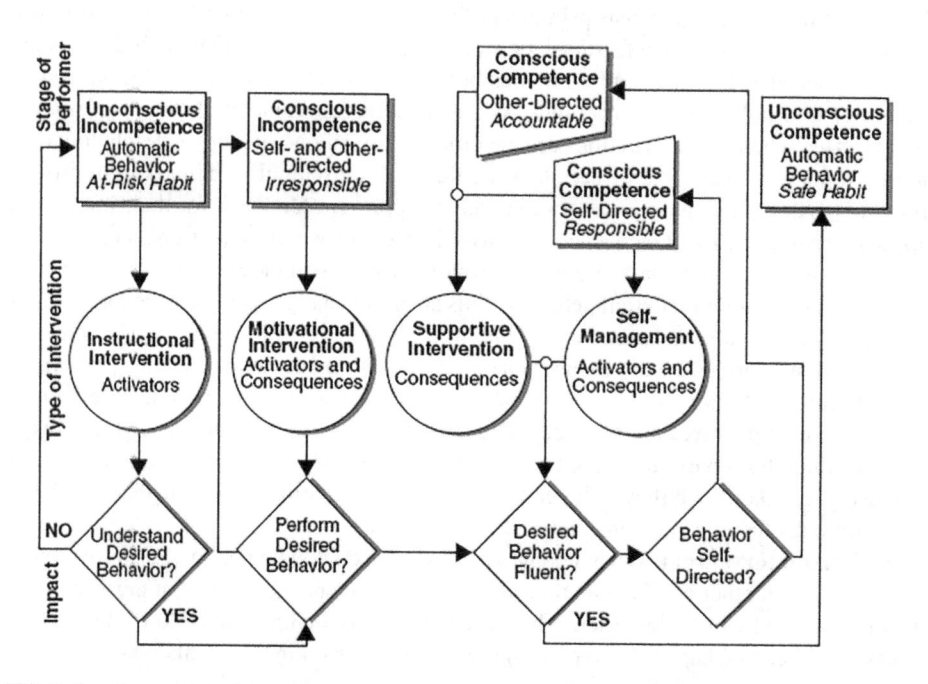

FIGURE 9.6 Awareness (conscious vs. unconscious) and safety-related behavior (competence vs. incompetence) determine which of four types of intervention is relevant.

Behavioral Safety Analysis

intervention is needed to get the behavior to a fluent state. Techniques for giving supportive recognition are described in Chapter 12.

Many people need a supportive consequence for their safe behavior. In other words, most experienced workers know what to do in order to prevent injury on their jobs, and they have performed their jobs safely one or more times, but the safe way might not be habitual. The individual is "consciously competent" but needs supportive recognition or feedback for response maintenance and increased fluency.

Figure 9.6 illustrates a distinction between conscious competence/other-directed and conscious competence/self-directed. If a safe work practice is self-directed, the employee is considered responsible and a self-management intervention is relevant. As detailed elsewhere (Geller, 2016; Watson & Tharp, 1997), the methods and tools of effective self-management are derived from applied behavioral science and are perfectly consistent with the principles of behavior-based safety.

In essence, self-management involves the application of the DO IT process introduced in Chapter 8 to one's own behavior. This means:

1. Defining one or more target behavior(s) to improve.
2. Monitoring those behaviors.
3. Manipulating relevant activators and consequences to increase occurrences of desired behavior and/or to decrease occurrences of undesired behavior.
4. Tracking continual change in the target behavior(s) in order to determine the impact of the self-management process.

9.3.4 Accountability vs. Responsibility

From the perspective of large-scale OSH success, the distinction in Figure 9.6 between accountable and responsible is critical. People often use the words accountability and responsibility interchangeably. Whether you hold someone accountable or responsible for getting something done, you mean the same thing. You want that person to accomplish a certain task and you intend on making sure it happens. However, let's consider the receiving end of this situation. How does a person feel about an assignment—does s/he feel accountable or responsible? Here is where a difference is evident and useful.

When you are held accountable, you are asked to reach a certain objective or goal, often within a designated time period. However, you might not feel responsible to meet the deadline, or you might feel responsible enough to complete the assignment, but that is all. You do only what is required and no more. In this case, accountability is the same as responsibility.

There are times, however, when you extend your responsibility beyond accountability. You do more than what is required. You go beyond the call of duty as defined by a particular accountability system. This is often essential when it comes to OSH. To improve safety beyond the current performance plateau experienced by many companies, workers need to extend their responsibility for safety beyond that for which they are held accountable. They need to transition from an other-directed state to a self-directed state.

Many jobs are accomplished by a lone worker. There is no supervisor or coworker available to hold the employee accountable for performing the job safely. The challenge for safety professionals and corporate leaders is to build the kind of work culture that enables or facilitates responsibility or self-accountability for safety. Practical strategies for making this happen are covered in Part 5, especially Chapter 16.

9.4 IN SUMMARY

This chapter offered some basic guidelines for diagnosing the human-behavior dynamics of OSH-related issues. Many situational, social, and individual factors contribute to a behavioral discrepancy—a distinction between the behavior performed and the behavior desired. In safety terms, this is the difference between at-risk and safe behavior.

Many of the factors contributing to a behavioral discrepancy are due to the context in which the task is performed or characteristics of the task itself. Common contextual variables that can influence a behavioral discrepancy are:

1. Unclear or misunderstood expectancies.
2. Upside-down contingencies with positive consequences for at-risk behavior or negative consequences for safe behavior.
3. The lack of behavior-based feedback to help people improve.

Often a job can be simplified or re-engineered to reduce physical or mental effort, which decreases the probability of personal injury.

Training should be considered only after critical contextual and task variables have been analyzed and corrected. It is usually optimal to include some education with the training—meaning relevant theory, principles, and rationale are presented to justify the step-by-step procedures taught and practiced during training. Adequate education also enables worker customization of procedures to fit a particular work context. This, in turn, leads to employee ownership of the process, feelings of responsibility, and increased engagement.

Education and training reflect the instructional approach to corrective action. This type of intervention is obviously most effective when the participants are willing to learn. They are initially unaware of the correct procedures and are therefore "unconsciously incompetent." Instruction will not help much for people who know what to do but do not do it. Those individuals are "consciously incompetent" and need a motivational intervention, as discussed later in Chapter 12.

For most employees, the issue is not a matter of knowing what behaviors are safe. They periodically perform all of the safe operating procedures called for on the job. The problem is consistency or fluency. They do not follow the safe protocol every time. These people need supportive intervention (e.g., behavioral recognition and/or gratitude) to keep them safe.

When safe work practices are relatively convenient, like putting on PPE or buckling a safety belt, the behavior can become habitual. When such behavior becomes a natural part of the work routine, the participant is considered "unconsciously competent." However, some behaviors, like locking out a power source, are relatively complex and never reach the automatic stage. Regular supportive intervention is often needed to keep such inconvenient OSH behaviors going, unless the individual is self-directed with regard to the particular behavior.

Self-directed individuals hold themselves accountable for doing the right thing, even when the behavior is relatively uncomfortable and inconvenient. These people certainly appreciate supportive intervention from managers, friends, and coworkers, but they keep performing the safe behavior when no one is around to hold them accountable. These self-directed or self-motivated workers hold themselves accountable. They feel responsible and go beyond the call of duty to prevent injuries to themselves and others. We call this "actively caring for safety," which is the focus throughout Part 5 of this book.

Part Four

Behavior-Based Intervention

10 Intervening with Activators

Intervention techniques to increase occurrences of safe behaviors or decrease occurrences of at-risk behaviors are either activators or consequences. This chapter explains activators, with real-world examples showing how to develop effective strategies. This discussion is framed by six principles for maximizing the impact of behavior-focused activators.

> Best efforts are not enough, you have to know what to do.
>
> **(W. Edwards Deming)**

Chapter 9 explained how the activator–behavior–consequence (ABC) model can be used to diagnose the contributing factors to an incident or at-risk behavior and to develop a plan for corrective action. With this chapter, we begin our discussion of intervention design and implementation to improve OSH-related behavior. As such, the ABC model is used as introduced in Chapter 8—a framework for designing behavior-change interventions.

First, let's realize the need for OSH interventions, because even maintaining our own safety-related behavior is not easy. It is usually a continuous fight with human nature, because in most situations activators and consequences naturally support risky behavior in lieu of safe behavior. At-risk behavior often allows for more immediate fun, comfort, and convenience than does safe behavior, inspiring the need for an intervention to direct and motivate occurrences of safe behavior. Activators are generally much easier and less expensive to use than consequences, so it is not surprising that they are employed much more often to promote safe behavior. Posters or signs are perhaps the most popular activator for safety.

- Some bear only a general message—"Safety is a Condition of Employment;" others refer to a specific behavior—"Hard Hat Required in this Area."
- Some signs request the occurrence of a behavior—"Walk," "Wear Ear Plugs in This Area"; others direct the avoidance of a certain behavior—"Don't Walk," "No Smoking Area."
- Sometimes a relatively convenient response is requested—"Buckle-Up," while other signs prompt relatively inconvenient behaviors—"Lock Out All Energy Sources Before Repairing Equipment."
- Some signs imply consequences—"Use Eye Protection: Don't Be Blinded by the Light"; others do not—"Wear Safety Goggles."
- We might be reminded of a general purpose—"Actively Care for Safety"—or a challenge—"100 Percent Safe Behavior is Our Goal This Year."

I have visited a number of work environments where all of these types of safety signs were displayed. In fact, I have seen situations that make the illustration in Figure 10.1 seem not very farfetched. Does this sort of "overkill" work to increase occurrences of safe behavior and reduce injuries? If you answered "No," then this time your commonsense was correct, because you have been there and experienced the ineffectiveness of many safety signs.

Which signs would you eliminate from Figure 10.1? How would you change certain signs to increase their impact? What activator strategy would you use instead of a sign? This chapter will enable you to answer these questions—not on the basis of commonsense but from applied behavioral science. Let's consider six key principles for increasing the impact of an activator.

FIGURE 10.1 Safety activators can be overwhelming and ineffective.

10.1 PRINCIPLE #1: SPECIFY THE DESIRED BEHAVIOR

Figure 10.2 illustrates "explosively" the need to include sufficient response information with a behavioral request, but too much specificity can bury a message, as illustrated in Figure 10.3. Activators ought to specify a desired response, but not overwhelm with complexity, as I have seen in a number of industrial signs. Overly complex signs are easy to overlook; over time they just blend into the woodwork. Keeping signs salient or noticeable is clearly a challenge.

FIGURE 10.2 Some activators are not specific enough.

Intervening with Activators

FIGURE 10.3 Some signs are too complex to be effective.

10.2 PRINCIPLE #2: MAINTAIN SALIENCE WITH NOVELTY

10.2.1 Habituation

It is perfectly natural for activators like sign messages to lose their impact over time. This process is called habituation, and it is considered by some psychologists to be the simplest form of learning.

Habituation happens even among organisms with primitive nervous systems. For example, when you lightly tap the shell of a large snail, it withdraws into its shell. After about 30 seconds the snail will extend its body from the shell and continue on its way. When you tap the shell again, the snail will withdraw again. However, this time the snail will stay inside its shell for a shorter duration. Your third tap will cause withdrawal again, but the withdrawal time will be even shorter. Each tap on the snail's shell results in successively shorter withdrawal time until eventually the snail will stop responding to your tap. The snail's behavior of withdrawal to the activator—shell tapping—will have habituated.

Habituation is perfectly consistent with an evolutionary perspective. If there is no obvious consequence (good or bad) from responding to a stimulus, the organism, whether an employee or a snail, stops reacting to it. It is a waste of time and energy to continue responding to an activator that seems to be insignificant. What would a snail do in a rainstorm if it did not learn to ignore shell taps that have no consequence?

Consider the distractions and distress you would experience daily if you could not learn to ignore noises from voices, radios, traffic, and machinery. At first, those environmental sounds might be quite noticeable and perhaps distracting, but through habituation they become insignificant background noise. They no longer divert attention nor interfere with ongoing performance.

What is the relevance of habituation for safety? It is human nature to habituate to everyday activators in our environment that are not supported by consequences. This is the case with many safety activators. Staying attentive to safety activators is a continuous fight with one feature of human nature—habituation.

10.2.2 Warning Beepers: A Common Work Example

Figure 10.4 illustrates quite clearly the phenomenon of habituation and reduced activator salience with experience. I bet you can reflect on personal experiences quite similar to the one shown here. Not only has the brick mason habituated to the familiar "beep" of the backing vehicle, but the driver is illustrating risk compensation (or risk homeostasis) as introduced in Chapter 6. He is not looking over his shoulder to check for a potential collision victim. He assumes the warning beeper is sufficient to activate coworkers' avoidance behavior and prevent an injury.

FIGURE 10.4 Some signals we rely on lose impact over time.

This particular activator has actually reduced the driver's perceived risk, which influences his at-risk behavior of looking forward instead of turning his head to check his rear. A key point: Understanding the basic learning phenomenon of habituation can prevent overreliance on activators and support a need to work more defensively.

10.3 PRINCIPLE #3: VARY THE MESSAGE

What does habituation tell us about the design of safety activators? Essentially, we need to vary the message. When an activator changes it can become more salient and noticeable. The "safety share" discussed in Chapter 7 follows this principle. When participants in a group meeting are asked to share something they have done for safety since the last meeting, the examples will vary considerably. Similarly, group discussions of close calls and potential corrective actions will also vary dramatically. The messages from safety shares and close-call discussions are also salient because they are personal, genuine, meaningful, and relevant.

10.3.1 Changeable Signs

Over the years I have noticed a variety of techniques for changing the message on safety signs. Some have removable slats to place different messages. Most readers have seen computer-generated signs with an infinite variety of safety messages. Some plants even have video screens in break rooms, lunch rooms, visitor lounges, and hallways that display many kinds of OSH-related messages, conveniently controlled by user-friendly computer software.

Who determines the content of those messages? You know who should—the target audience for those signs. The same people expected to follow the specific behavioral advice should have as much input as possible in defining message content. Many organizations can get suggestions for safety messages just by asking. However, if employees are not accustomed to giving safety suggestions, they might need a positive consequence to motivate their input.

10.3.2 WORKER-DESIGNED SAFETY SLOGANS

In 1985, employees and visitors driving into the main parking lot for Ford World Headquarters in Dearborn, MI, passed a series of four signs arranged with sequential messages, like the old Burma Shave signs. The messages were rotated periodically from a pool of 55 employee entries in a limerick contest for safety-belt promotion. My three favorites are illustrated in Figure 10.5.

Notice that the last sign in each series of safety-belt promotion messages at Ford World Headquarters includes the name and department of the author. Such public recognition, with the author's permission of course, provided a positive consequence or reward for the participants. It also reminded all sign viewers that many different people from various work areas were actively involved in OSH. Through positive recognition and observational learning, including vicarious reinforcement, this simple technique promoted ownership and involvement in an OSH process. This leads to the next principle—involve the target audience.

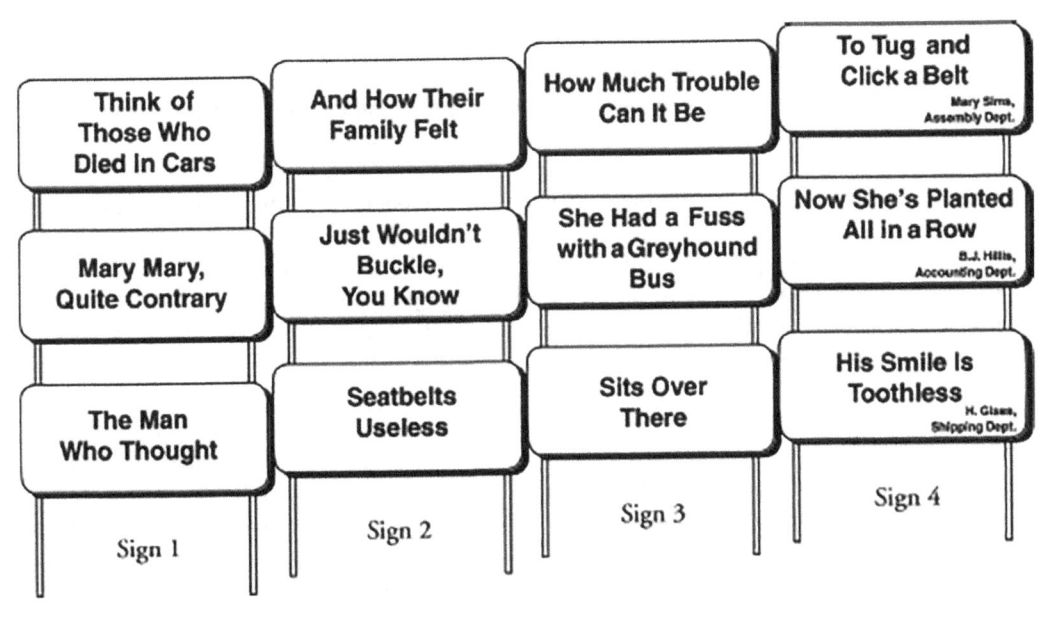

FIGURE 10.5 In 1986, Ford employees created buckle-up activators for display at Ford World Headquarters.

10.4 PRINCIPLE #4: INVOLVE THE TARGET AUDIENCE

Figure 10.6 depicts a sample promise card for involving people in a commitment to perform a particular behavior. The behavior targeted to increase in frequency could be selected by a safety director or group leader, or through a group-consensus discussion. This behavior is written on the promise card, and group members decide on the duration of the promise period and write the end date on the card. Then, each group member should be encouraged, not coerced, to sign and date

Safe Behavior Promise Card

I promise to _____

From _____ until _____
                           date
signature

FIGURE 10.6 A promise card activates a behavioral commitment.

a card. This group application of the safe behavior promise strengthens a sense of group cohesion or belonging. Follow these procedural points for optimal results:

- Define the desired target behavior specifically.
- Involve the group in discussing the personal and group value of the target behavior.
- Make the commitment for a specified period of time that is challenging but not overwhelming.
- Assure everyone that signing the card is only a personal commitment, not a company contract.
- There should be no penalties (not even criticism) for breaking a promise.
- Encourage everyone to sign the card, but do not use pressure tactics.
- Signers should keep their promise cards in their possession or post them in their work areas as reminders.

The more involvement and personal choice solicited during the completion of this activator strategy, the better each individual will feel about the process. Personal commitment to perform a certain behavior is activated as a result; those involved in the process should feel obligated to fulfill the promise. Signing the card publicly in a group meeting also implicates social consequences to motivate compliance. That is, many participants will be motivated to keep their promise to avoid disapproval from a group member. When individuals keep their promise, recognition and approval from the group supports and rewards maintenance of the targeted safe behavior.

10.4.1 The "Flash for Life"

This activator intervention dates back to 1984, when I developed the "Flash for Life." Here is how it works. A person displays to vehicle occupants the front side of an 11- by 14-inch flash card that reads, "Please Buckle-up—I Care." If someone buckles up after viewing this message, the "flasher" flips the card over to display the bold words, "Thank You for Buckling Up."

For the first evaluation of this behavior-change intervention, the "flasher" was in the front seat of a stopped vehicle and the "flash recipient" was the driver of an adjacent, stopped vehicle. The flash card was shown to 1087 unbuckled drivers, and of the 82 percent who looked at the card, 22 percent complied immediately with the buckle-up request (Geller et al., 1985). My youngest daughter, Karly, was the "flasher" for about 30 percent of the trials in that study. On a few occasions we got a hand signal that was not used to indicate a right or left turn. Once Karly asked: "Daddy, what does that mean?" and I answered, "It means you're number one Honey; they're just using the wrong finger."

When hearing about this "Flash for Life" project, many of my colleagues expressed concern for my sanity. "Why do you waste your time?" some would say. "Getting 22 percent to buckle up is not a big deal, and most of those who buckled up for your daughter only did it that one time. They probably won't buckle up the next day."

I had two answers to such pessimism. First, achievement is built on "small wins." People need to break up big problems or challenges into small, achievable steps and then work on each successive step, one at a time. We cannot expect to solve a major safety problem like low use of PPE with one intervention technique, but we need to start somewhere. If everyone contributed a "small win" for OSH, the cumulative effects could be tremendous.

My second reply focused on the powerful influence of involvement. This intervention procedure enabled my young daughter to get involved in a safety project, even though she did not yet understand the concept of "safety." Every time Karly "flashed" another person to buckle up, her own commitment to practice the target behavior increased. I never had to remind Karly to use her safety belt.

10.4.2 The Airline Lifesaver

Another personal experience with an activator intervention is one I started using in November 1984, and I continued to use this activator strategy for 17 consecutive years. Whenever boarding a commercial airplane, I handed the flight attendant a 3- by 5-inch "Airline Lifesaver" card. The card indicated that airlines have been the most effective promoters of seat-belt use and requested that someone in the flight crew make the following announcement: *Now that you have worn a seat belt for the safest part of your trip; the flight crew would like to remind you to buckle up during your ground transportation.*

From November 1984 until May 2001, I distributed the "Airline Lifesaver" on 1050 flights, and on more than 40 percent of those occasions, the flight attendant gave a public buckle-up reminder.

Many colleagues laughed at the "Airline Lifesaver," claiming I was wasting my time. A common comment was, "No one listens to the airline announcements anyway, and besides, do you really think an airline message could be enough to motivate people to buckle up if they don't already?"

Please consider this personal experience from the mid-1980s. I observed a woman approach the driver of an airport shuttle, and ask her to "Please use your safety belt." The driver immediately buckled up. When I thanked the woman for making the buckle-up request, she replied that she normally would not be so assertive but she had just heard a buckle-up reminder on her flight, "and if a stewardess can request safety-belt use, so can I."

Except for a few anecdotes like this one, it is impossible to assess the direct buckle-up influence of the Airline Lifesaver. However, it is "safe" to assume that the beneficial, large-scale impact of this activator is a direct function of the number of individuals who deliver the reminder card to airline personnel. If the delivery of an Airline Lifesaver does not influence a single airline passenger to use a safety belt during ground transportation, at least the act of handing an Airline Lifesaver card to another person should increase the card deliverer's commitment to personal safety-belt use.

Of course, the primary purpose of getting involved in a safety intervention is to prevent injury or improve a person's quality of life. Unfortunately, we rarely see such important consequences.

Thus, we need self-motivation, feedback, interpersonal approval, and self-talk. We tell ourselves the safe behavior is "the right thing to do," and that someday an injury will be prevented. We cannot count the number of injuries we prevent; we just need to "keep the faith."

On December 28, 1994, I received a special letter from Steven Boydston, then assistant vice president of Alexander & Alexander of Texas, Inc., which helps me "keep the faith" that the "Airline Lifesaver" made a difference. The encouraging words in that letter are repeated in Figure 10.7. This success story is itself an activator for such proactive interventions as the Airline Lifesaver. It sure worked for me.

THANKS!!! On December 11, 1994 I was a passenger on Flt. 499 from Houston to San Francisco. At the end of the flight the pilot came on the speaker and said: "Now that the safest part of your journey is over and you are about to make the most dangerous part of your journey, please remember to use your seat belt...." I may have the words a little off, but I think you know the message. This was the second time I heard the message. The first was at the ASSE PDC in Dallas.

I am an obsessive seat belt user but when I got into the taxi, the seat belt was, as usual, buried in the seat. Usually when I find this in a cab I say, "the heck with it," but I can honestly say the pilot's message motivated me and I "dug out" the seat belt.

At over 70 mph the taxi hydroplaned and struck the guardrail. Thank God for the new barriers that prevent cars from being thrown back onto traffic. Thank you and the pilot for the reminder. My wife and children are also grateful. I suffered a neck and shoulder injury – I think it is relatively minor. It could have been much worse. I should mention the driver was O.K. He was wearing his seat belt.

In safety it is seldom we can point to a particular event we can take credit for. This is one for you. Thanks again.

FIGURE 10.7 I received these words of encouragement from Steven Boydston on December 28, 1994.

10.5   PRINCIPLE #5: ACTIVATE CLOSE TO RESPONSE OPPORTUNITY

Most of the effective activators discussed so far occurred at the time and place the target behavior should happen. The "Flash for Life" card was presented when people were in their vehicles and could readily buckle up, and I gave airline attendants the "Airline Lifesaver" card when boarding the plane. Actually, I believe I would have received more compliance with the request for a buckle-up announcement if I had handed an attendant the announcement card at the end of the flight—closer to the opportunity to make the requested response. In fact, when I inquired about the lack of a buckle-up announcement while deplaning, the most common excuse was "I forgot."

10.6   PRINCIPLE #6: IMPLICATE CONSEQUENCES

Many of the successful activator strategies illustrated in this chapter were explicitly or implicitly connected to consequences. Signing a promise card or public declaration, for example, implicates social approval vs. disapproval for honoring vs. disavowing a commitment. Positive recognition has motivated employees to create safety slogans, and the most influential activators usually made reference to consequences. Relatedly, the salient beep of a radar detector effectively

Intervening with Activators

FIGURE 10.8 Negative consequences can increase the subsequent impact of an activator.

motivates reduced vehicle speeds because it enables drivers to avoid a negative consequence—an encounter with a police officer.

Figure 10.8 illustrates the influence of negative consequences on the impact of an activator. In this case, however, the compliance will be reactive rather than proactive. That is, a negative incident occurred because the specific behavior-focused instructions were not followed. From now on, however, it is likely that activator will be effective for this person. And if s/he shares the negative incident and its messy consequences with other store personnel, this activator will take on increased significance and behavioral impact.

10.6.1 INCENTIVES VS. DISINCENTIVES

Activators that signal the availability of a consequence are either incentives or disincentives. An *incentive* announces to an individual or group, in written or oral form, the availability of a reward. That positive consequence follows the occurrence of a certain behavior or the outcome of one or more behaviors. In contrast, a *disincentive* is an activator announcing or signaling the possibility of receiving a penalty. That unpleasant consequence is contingent on the occurrence of a particular undesirable behavior.

The next chapter discusses how to design and apply consequences to motivate behavior. At this point, it is important to understand that the *power of an activator to motivate behavior depends on the consequence(s) it signals.* Figure 10.9 illustrates this connection between activator and consequence. If a sign like the one shown in Figure 10.9 motivated a driver to attempt safer driving practices, it would work because of the potential consequences implied by the activator. Every time the driver got into the vehicle, she would be reminded of potential consequences for certain driving practices. Incidentally, do you perceive the sign on the vehicle in Figure 10.9 as an incentive or a disincentive?

My guess is you perceived it as a disincentive rather than an incentive. I bet you saw the sign as a threat to reduce at-risk driving rather than an incentive to encourage safe driving. You would

FIGURE 10.9 The most powerful activators imply immediate consequences.

not expect Dad to get a phone call commending his daughter's driving. If any phone calls were made, they would be to report at-risk or discourteous driving. That is exactly how my daughter, Krista, perceived the activator in Figure 10.9. As the illustration suggests, she flatly refused to drive with such a sign—a perceived disincentive—on the back of our car.

At the time of this writing, a widely recognized, public safety slogan is, "If you see something, say something," which most people interpret as encouragement "to report suspicious, or potentially dangerous activities to the authorities . . . to help prevent or mitigate potential risks and contribute to maintaining a safe environment for everyone" (from ChatGPT, 7/17/2013). In a recent commentary (Geller, 2023), I suggested a paradigm shift by looking for desirable behavior and situations to appreciate, and then providing positive recognition or supportive feedback and sincere gratitude to the individual(s) responsible for the desirable behavior and/or situation.

Then, I asked the readers to imagine a work, school, or home culture in which desirable behavior gets more attention, recognition, and support than undesirable behavior. In such a culture, desirable behavior would become the norm, and such undesirable behavior as interpersonal conflict and violence would become increasingly rare. Why? Because the best way to eliminate undesirable behavior is to recognize and support incompatible desirable behavior.

10.6.2 Setting Goals for Consequences

Let's talk about safety goals in the context of activators that imply consequences. Dr. Deming advised us to "eliminate slogans, exhortations, and targets for the workforce . . . eliminate work standards . . . management by objectives, and management by the numbers" (Deming, 1985). Does that mean we should stop setting safety objectives and goals? Should we stop trying to activate safe behaviors with signs, slogans, and goal statements? Does this mean we should stop counting OSHA recordables and lost-time cases, and stop holding people accountable for their work injuries?

Answers to all of those questions are "Yes," if you take Deming's points literally. However, my evaluation of Deming's scholarship and workshop presentations, and my personal communications with him in 1990 and 1991, have led me to believe that Dr. Deming meant we should

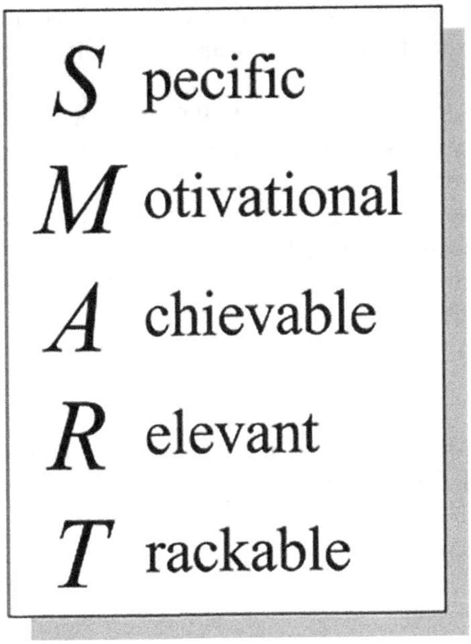

S pecific
M otivational
A chievable
R elevant
T rackable

FIGURE 10.10 SMART goals are effective activators.

eliminate goal-setting, slogans, and work targets as they are currently implemented. Dr. Deming was not criticizing the appropriate use of goal-setting, management by objectives, and activators; rather he was lamenting the frequent incorrect use of these activator interventions.

Set SMART goals. I remember the techniques for setting effective goals with the acronym SMART, as illustrated in Figure 10.10. SMART goal-setting defines what will happen when the goal is reached (the consequences), and tracks progress toward achieving the goal. Rewarding feedback for completing intermediate steps toward achieving the ultimate goal is a consequence that motivates continued progress. Of course, it is critical that the people asked to work toward the goal "buy in" or believe in the goal. They must believe the goal is relevant for achieving a worthwhile consequence, and that they have the skills and resources to achieve it.

Many readers will recognize the SMART goal acronym, because of the popular audio tape created many years ago by Zig Ziglar. Whenever I ask audiences the relevance of "M" for SMART goals, several participants inevitably shout out "Measurable," because that is from the Zig Ziglar version. Please note my substitution of "Motivational" for "Measurable." Why? Because it is essential to be motivated to achieve a particular goal, and this happens after you specify the positive consequence(s) attained when reaching that goal. Yes, it is also crucial to measure progress toward goal achievement, and therefore my "T" for SMART goals reflects "Trackable" rather than the "Timely" in Zig Ziglar's version of SMART goal-setting.

Focus on the process. Safety goals should focus on process activities that can contribute to injury prevention. Workers need to discuss what they can do to reduce injuries—from reporting and analyzing close calls to conducting safety audits of environmental conditions and work practices. One safety steering committee I worked with wanted to increase daily interpersonal communications regarding safety. They set a goal for their group to achieve 500 safety communications within the following month. To do this, they had to develop a system for tracking and recording interpersonal "safety talks." They designed a wallet-sized "SMART Card" for recording their interactions with others about safety. One member of the group volunteered to tally and graph the daily card totals.

Another work group with whom I consulted set a goal of 300 behavioral observations of lifting. Employees had agreed to observe each other's lifting behaviors according to a critical behavior checklist they had developed. If each worker completed an average of one lifting observation per day, the group would reach their goal within the month. Each of those work groups reached their OSH goals within the expected time period, and as a result they celebrated their "small win" at a group luncheon.

These two examples illustrate the use of SMART goals and depict OSH as process focused and achievement oriented, rather than the more common and less effective outcome-focused and failure-oriented approach promoted by injury-focused goals. More importantly, those goals were employee driven. Workers were motivated to initiate the safety process because it was their idea. They got involved in the process and owned it. They stayed motivated because the SMART goals were like a roadmap telling them where they were going, when they would get there, and how to follow their progress along the way.

10.7 IN SUMMARY

This chapter presented examples of intervention techniques called activators. Activators occur before desired or undesired behavior in order to direct behavior from potential performers. Based on rigorous behavioral science research and backed by real-world examples, six principles for maximizing the impact of activators were explained:

- Specify behavior.
- Maintain salience with novelty.
- Vary the message.
- Involve the target audience.
- Activate close to response opportunity.
- Implicate consequences.

We are constantly bombarded with activators. At home we get telephone solicitations, junk mail, television commercials, and verbal requests from family members. At work, it is phone mail, e-mail, memos, policy pronouncements, and verbal directions from supervisors and coworkers. On the road, there is no escape from billboards, traffic signals, vehicle displays, radio ads, and verbal communication from people inside and outside our vehicles. As discussed in Chapter 5, we selectively attend to some of those activators and ignore others. Only a portion of the activators we perceive actually influence our behavior. Understanding the six principles discussed in this chapter can help predict which activators will influence behavior.

Obviously, we do not need more activators in our lives. We certainly do need *more effective* activators to promote safety and health. It would be far better to make a few safety activators more powerful than to add more activators to an OSH system already overloaded with information. We need to plan our safety activators carefully so the important safety directives receive the attention and ultimate action they deserve.

If you want an activator to motivate action, you need to imply consequences. The most powerful activators make the observer aware of consequences available following the performance of a target behavior. Consequences can be positive or negative, intrinsic or extrinsic to the task, and internal or external to the person. The next chapter explains the preceding sentence, which is key to getting the most beneficial behavior from an intervention process.

11 Intervening with Consequences

Consequences motivate behavior and influence related attitudes. This happens in various ways. Consequences can be positive or negative, intrinsic (natural) or extrinsic (extra) to a task, and internal or external to a person. These characteristics need to be considered when designing and evaluating intervention programs. This chapter explains why and provides principles and practical procedures for motivating people to work safely over the long term. In other words, you will learn how to influence behavior and attitudes so that both are consistent with a Total Safety Culture.

> Every act you have ever performed since the day you were born was performed because you wanted something.
>
> **(Dale Carnegie)**

The introductory quotation from Dale Carnegie's classic book, *How to Win Friends and Influence People*, first published in 1936, represents a key principle of human motivation and BBS. Although supported by substantial research (Skinner, 1938), it actually runs counter to some commonsense.

Think about it. When people ask us why we did something, we are apt to say, "I wanted to do it," "I was told to do it," or "I needed to do it." Those explanations sound as if the cause of our behavior comes before we act. This perspective is supported by numerous "pop psychology" self-help books and audiotapes that say people motivate themselves with positive self-affirmations or optimistic thinking and enthusiastic expectations. In other words, behavior is presumed to be caused by some external request, order, or signal, or by an internal force, drive, desire, or need.

This chapter explains the fallacy in that commonsense and shows ways to maximize the impact of an extrinsic reward contingency. Again, the research-supported principle is that activators direct behavior and consequences motivate behavior, but the type of consequence certainly influences the amount of motivation, as this chapter elucidates.

11.1 THE POWER OF CONSEQUENCES

Popular author and humorist Robert Fulghum (1988) wrote *All I Really Need to Know I Learned in Kindergarten*, claiming he learned all the basic rules or norms for socially acceptable adult behavior as a young child. The list of rules in Figure 11.1 was excerpted from Fulghum's seminal book. Rules like "share everything," "play fair," "don't fight," and "clean up your own mess" were taught to most of us early on. Those are clearly ideal edicts to live by. Perhaps you still recall a teacher or parent using those rules to try to shape your behavior. Did it work? Do you follow each of those basic norms regularly, for no other reason or consequence except your realization that it is the right thing to do?

Imagine what a better world we would live in if everyone followed the simple rules listed in Figure 11.1 from a self-directed, principle-focused perspective. Alas, there are signs everywhere that this is just not the case.

The last two kindergarten rules in Figure 11.1 are directly relevant to OSH and in fact reflect basic themes of this text. As discussed previously, OSH requires people to stick together in a

❏ Share everything
❏ Play fair
❏ Don't hit others
❏ Put things back where you found them
❏ Clean up your own mess
❏ Don't take things that aren't yours
❏ Say you're sorry when you hurt someone
❏ Wash your hands before you eat
❏ Flush
❏ When you go out into the world, watch out for traffic, hold hands, and stick together
❏ And remember the Dick-and-Jane books and the first word you learned--the biggest word of all: **LOOK**

FIGURE 11.1 Basic rules of social life we learned well as children we do not necessarily follow as adults. (Excerpted from Fulghum, 1988)

spirit of shared belongingness and interdependence. However, sometimes we need activators to remind us of this critical rule and consequences to keep us working together for OSH.

"LOOK," Fulghum's last rule, is key to BBS and to achieving a TSC. This implies the "defensive working style" employees need to adopt. In a TSC, everyone looks for ways to improve OSH by intervening to reduce the occurrence of at-risk behaviors and increase the occurrence of safe behaviors. Chapter 10 exemplified ways to intervene with activators. Here, we focus on the more powerful intervention strategy—manipulating consequences.

11.1.1 INTRINSIC VS. EXTRINSIC CONSEQUENCES

Most applied behavioral scientists view "intrinsic motivation" differently from the description used in pop psychology books. The behavioral science perspective of intrinsic motivation is supported by research and our everyday experience. Plus, it is objective, practical, and useful for developing situations and programs to motivate behavior change.

Simply put, "intrinsic" does not mean "inside" people, where it cannot be observed, measured, and directly influenced. Rather, "intrinsic" refers to the nature of the task in which an individual is engaged. Intrinsically motivated tasks or behaviors, lead *naturally* to external consequences that support the behavior (supportive feedback) or give information useful for improving the behavior (corrective feedback).

Most athletic performance, for example, includes natural or intrinsic consequences that give supporting or correcting feedback. These consequences, intrinsic to the task, tell us immediately how well we have performed at swinging a golf club, shooting a basketball, or casting a fishing lure, for example. They motivate us to keep trying, sometimes after adjusting our behavior as a result of the natural feedback directly related, or intrinsic to the task.

Take a look at the fisherman in Figure 11.2. Some psychologists would claim he is motivated from within, or internally motivated. They use the term "intrinsic motivation" to refer to this person-state. In contrast, behavioral scientists consider the external consequences that naturally

FIGURE 11.2 Some tasks are naturally motivating because of intrinsic consequences.

motivate the fisherman's behavior. These cause him to focus so completely on the task at hand that he is not aware of his wife's mounting anger—or he is simply ignoring her. He may also be unaware that his supply of fish is creating a potential hazard. Similarly, OSH can be compromised because of excessive motivation for production. Rewards that are intrinsic to production can cause such motivation.

Notice that the "worker" in this picture does not receive a reward for every cast. In fact, he is on an intermittent reinforcement schedule. He catches a fish once in a while. That kind of reward schedule is most powerful in maintaining continuous behavior. Anyone who has gambled understands. Some say gambling is a disease, when in fact gambling is behavior maintained by intermittent rewarding consequences.

Some tasks do not provide intrinsic or natural feedback. In this case, it is often necessary to add an extrinsic, or extra, consequence to support or redirect the behavior. Many, if not most, safety-related behaviors fall in this category.

In fact, many safety practices have intrinsic negative consequences, such as discomfort, inconvenience, and reduced pace that naturally discourage their occurrence. Thus, to shape and maintain the occurrence of safe behaviors there is often a need for extrinsic supportive consequences, like intermittent praise, recognition, novelties, and credits redeemable for prizes. The intent is not to control people, but to help people control their own behavior by offering positive reasons for making safe choices.

Now look at the student in Figure 11.3. He expects an extrinsic positive consequence for completing an accurate calculation. Do you see a problem here? Sure, the pupil should feel good about giving the right answer. In other words, the intrinsic consequence of completing a task correctly should be perceived as valuable and rewarding by the student. The student should perceive the important payoff as getting the right answer.

Now we are talking about a person's interpretation of the situation, which is referred to as "internal" consequences in the next section. First, let's understand a very important point reflected in Figure 11.3. Whenever there is an observable intrinsic consequence to a task, the

FIGURE 11.3 External rewards can reduce internal motivation.

instructor, supervisor, or safety coach needs to help the performer see that consequence and realize its importance. In other words, we need to help people perceive the intrinsic consequences of their performance and show appreciation and pride in that outcome. This helps to make the intrinsic consequence rewarding for the performer, thereby facilitating ongoing motivation. So, if the teacher in Figure 11.3 displayed genuine approval and delight in the student's achievement, an extra extrinsic reward might not be needed to keep the student motivated.

11.1.2 Internal vs. External Consequences

The intrinsic and extrinsic consequences discussed so far are external to the performer. In other words, they can be observed by another person. Behavioral scientists focus on these types of consequences because they can be objective and scientific when dealing with external, observable aspects of people.

Behavioral scientists, however, do not deny the existence of internal factors that motivate action. There is no doubt that we talk to ourselves before and after our behaviors, and that self-talk influences our performance. We often give ourselves internal verbal instructions, called intentions, before performing certain behaviors. After our activities, we often evaluate our performance with internal consequences. In the process, we might motivate ourselves to press on (with self-commendation) or to stop (with self-condemnation).

When it comes to OSH, internal consequences to support the right behavior are critical. Remember, external and intrinsic (natural) consequences for safe behaviors are not readily available, and we cannot expect to receive sufficient support (extra consequences) from others to sustain our proactive, safe, and healthy choices. So, we need to talk to ourselves with sincere conviction to boost our OSH intentions. We also need to give ourselves genuine self-recognition after we do the right thing to keep ourselves going. When we receive special external supportive consequences from others for our efforts, we need to savor those and use them later to boost our self-motivation.

Intervening with Consequences

FIGURE 11.4 Some tasks require supportive consequences.

11.2 MANAGING CONSEQUENCES FOR OSH

At this point, I am sure you appreciate the special message reflected in Figure 11.4. Submitting safety suggestions is an activity not typically followed by external motivating consequences. In many work cultures, the idea of safety suggestions has long since passed. The suggestion boxes are empty. Does this mean there are no more good suggestions? Is the workforce not creative enough? You know the answer to both of those questions is a resounding "No."

Consider this example. I once worked with safety leaders at a Toyota Motor manufacturing plant in Georgetown, Kentucky, whose 6,000 employees submitted more than 35,000 production or safety-related suggestions in 1994. A greater number of suggestions were expected in 1995. Many employees in this culture were motivated internally to submit suggestions, but external consequences were in place to keep the process going.

Employees received timely feedback regarding the utility and the feasibility of every suggestion, and if the suggestion was approved, they were empowered to implement it themselves. Also, the individual or team responsible for the suggestion and application received 10 percent of the monetary savings that resulted from the implemented suggestion for the first year. Such external, extra and meaningful—in this case economic—consequences motivated a large workforce to make a difference.

11.2.1 THE CASE AGAINST NEGATIVE CONSEQUENCES

To subdue factors that influence at-risk behavior, it is often tempting to use a negative consequence or a penalty. All that is needed is a policy statement or some type of top-down mandate specifying a soon, certain, and sizable negative consequence following specific observable at-risk behavior. Could such a contingency be powerful enough to override the many natural positive consequences for taking risks?

FIGURE 11.5 The threat of a negative consequence is motivating.

Yes, behavioral scientists have found that negative consequences can permanently suppress behavior if the penalty is severe, certain, and immediate. However, before using "the stick," you should consider the following side-effects of using negative consequences to influence behavior.

Escape. Animals and people attempt to avoid situations with a predominance of negative consequences. Sometimes, this means staying away from those who administer the penalties. Humans will often attempt to escape from negative consequences by simply "tuning out" or perhaps cheating or lying. Of course, the ultimate escape from excessive negative consequences is suicide. Indeed, it is not uncommon for an individual to commit suicide in order to escape control by aversive stimulation, which can include the intractable pain of an incurable disease, physical or psychological abuse from a family member, or perceived harassment by an employee or coworker.

Unpleasant attitudes or emotional feelings are produced when people work to escape or avoid negative consequences. As shown in Figure 11.5, the threat of a negative consequence can influence behavior dramatically, but such situations are usually unpleasant for the "victim." Under fear-arousal conditions, people will be motivated to do the right thing, but only when they have to. They feel controlled, and, as discussed in Chapter 6, this can lead to distress and burnout. Obviously, that type of contingency and side effect is incompatible with a TSC where people feel "in control" and are ready and willing to go beyond the call of duty for another person's safety and health.

Aggression. Instead of escaping, people might choose to attack those perceived to be in charge. For example, murder in the workplace has increased in the U.S., and the most frequent cause appears to be reaction to or frustration with control by negative means. An aggressive reaction to this kind of control, however, might not be directed at the source.

An employee, frustrated by top-down aversive control at work, might not assault his or her boss directly but rather slow down production, sabotage an OSH program, steal supplies, or vandalize industrial property. Alternatively, the employee might react with spousal abuse at home. Then, the abused spouse might react by slapping a child. The child, in turn, might punch a younger sibling, and the younger sibling might punch a hole in a wall or kick the family pet—all as a result of perceived control by negative consequences.

Apathy. Apathy is a generalized suppression of behavior. In other words, negative consequences not only suppress the target behavior but they might also inhibit the occurrence of desirable behavior. Regarding OSH, this could mean a decrease in employee engagement. When people feel controlled by negative consequences, they are apt to simply resign themselves to doing only what is required. Going beyond the call of duty for a coworker's health or safety is out of the question.

Countercontrol. No one likes to feel controlled, and situations that influence these feelings in people do not encourage buy-in, commitment, and involvement. In fact, some people only follow top-down rules when they believe they could get caught, typified by drivers slowing down when noticing a police car. Some people look for ways to beat the system they feel is controlling them, like those vehicle drivers who purchase radar detectors. This is an example of "countercontrol," the fourth undesirable side-effect of negative-consequence contingencies.

I met an employee once who exerted countercontrol by wearing safety glasses without lenses. When wearing his "safety frames," he got attention and approval from various coworkers. Perhaps those coworkers were rewarded vicariously when seeing him beat the system they perceived was controlling them also.

Figure 11.6 illustrates an example of countercontrol. Although the supervisors might view the behavior as "feedback," it is countercontrol if it occurred to regain control or assert personal freedom. A perceived loss of control or freedom is most likely when a negative-consequence contingency is implemented. Also, countercontrol behavior is typically directed at those in control of the negative consequences.

FIGURE 11.6 Countercontrol is usually directed at those in charge of negative consequences.

11.2.2 Discipline and Involvement

How about traditional "discipline" for OSH—a form of top-down control with negative consequences? I have met many managers who include a "discipline session" as part of the corrective action for an injury report. The injured employee gets a negative lecture from a manager or supervisor whose safety record and personal performance appraisal were tarnished by the injury.

Such "discipline sessions" are unpleasant for both parties and certainly do not encourage personal commitment or buy-in to the safety mission of the company. Instead, the criticized and embarrassed employees are simply reminded of the top-down control aspects of corporate OSH, usually resulting in increased commitment *not* to volunteer for safety interventions, nor to encourage others to participate. In this case, the culture loses the involvement of invaluable OSH participants.

What about progressive discipline? Whenever I taught behavior-management principles and procedures, the question of how to deal with the repeat offender inevitably came up. Are there not times when punishment is necessary? Doesn't an individual who "willfully" breaks the rules after repeated warnings or confrontations deserve a penalty? Through "progressive discipline," these individuals receive top-down penalties, starting with a verbal warning, then written warnings, and eventually dismissal. In some cases, dismissal is the best solution for noncooperative individuals who can be a divisive and dangerous factor in the workforce. Fortunately, this worst-case scenario is rare.

The standard progressive discipline approach in safety enforcement includes three steps. After the third infraction, it is common to send the employee home for a certain number of days without pay. In other words, "three strikes and you're out." But the wrongdoer is not out for good. The individual is usually allowed back "in the game." Here is the crucial question. Is the person a better "player" upon his or her return?

When employees are penalized by being temporarily dismissed, we expect them to perform better when they return to work. In other words, we hope they learned something from that demeaning negative consequence. We also hope the learning is more than how to avoid getting caught next time.

Actually, whether the right or wrong kind of learning occurs in this situation depends on one key factor—attitude. If the employee is angry and does not own up to taking a calculated risk, useful learning is unlikely. If negative or hostile emotions develop in an employee as a result of the dismissal, do not expect that employee to return to work with a more pleasant and cooperative demeanor. Instead, expect a more disgruntled worker, who might give lip service to following the safety rules in order to avoid another dismissal, but s/he will likely share a negative attitude with anyone willing to listen. As we have all experienced, "returning a rotten apple to a barrel makes other apples it contacts rotten."

Positive discipline. One way to avoid this problem is not to send an employee home *without pay*. Instead, dismiss the employee *with pay*. Grote (1995) calls this "positive discipline." This is not about docking one's wages for a safety infraction. It is about finding a meaningful way to reduce a behavioral discrepancy. Now, this is not a free vacation day by any stretch of the imagination. The employee is required to think about the calculated risk and decide what can be done to eliminate that at-risk behavior. By not withholding wages, this evaluative process is not tainted by a negative or hostile attitude.

It should be clear that one option for the employee to consider is not to return to work. The individual should seriously consider whether s/he can meet the safety standards of the company. Is it too difficult to perform consistently with the paradigm of OSH as a core value?

It is unlikely a person will admit to not holding OSH as a core value. Thus, it is realistic and relevant for the employee to conduct a behavioral analysis (as outlined in Chapter 9) and then develop a personal corrective action plan for reducing the behavioral discrepancy implied by the rule infraction or calculated risk. Thus, at the end of the dismissal day(s), the ultimate deliverable is a specific list of things the employee will do to reduce the behavioral discrepancy and realign work practice with OSH as a value.

This corrective action plan should include a specification of environmental and interpersonal support the individual will summon in order to meet an improvement objective. For example, the employee might recommend a modification of a workstation to make the desired safe behavior more convenient or add an activator to the area as a behavioral reminder. The action plan might also include a solicitation of peer support by requesting that certain coworkers offer directive and/or supportive feedback (as detailed in Chapter 12).

It is critical for a supervisor or safety leader to review this corrective action plan as soon as the employee returns to work. Both parties must agree that the plan is reasonable, feasible, and cost effective. It is likely that mutual agreement and commitment to a suitable action plan will require significant discussion, consensus building, and refinement of the document. The final documentation of the plan should be signed by both parties. When a person signs a personal commitment to a document that took some effort to develop, the probability of compliance is greatly enhanced.

11.3 "DOS" AND "DON'TS" OF SAFETY REWARDS

Now let's look at the flip side of negative consequences—rewards. In one-on-one situations with children at home or in school, using positive consequences to increase desirable behavior is straightforward and easy. However, using rewarding consequences effectively with adults in work settings is easier said than done, especially when it comes to OSH. Throughout my 50+ years of professional experience in motivational psychology, I have seen more inappropriate reward programs in OSH than in any other area. This is unfortunate, because the effective use of extra positive consequences is often critically important to overcome the readily available factors that support at-risk behavior.

11.3.1 Doing It Wrong

Many incentive/reward programs for OSH do not specify behavior. Employees are rewarded for avoiding a work injury or for achieving a certain number of "safe work days." So, what behavior is incentivized? Right on—not to report injuries.

If having an injury loses one's reward, or worse, the reward for an entire work group, there is pressure to avoid reporting that injury, if possible. Many of these nonbehavioral, outcome-based incentive programs involve substantial peer pressure because they use a group-based contingency. That is, if anyone in the company or work group is injured, everyone loses the reward. Not surprisingly, I have seen coworkers cover for an injured employee in order to keep accumulating "safe days" and not lose their chance to receive a reward.

Those incentive programs might decrease the numbers of reported injuries, at least over the short term, but corporate safety is obviously not improved. Indeed, such programs often create apathy or helplessness regarding OSH achievement. Employees develop the perspective that they cannot really control their injury record but must cheat or beat the system to celebrate the "achievement" of an injury-reduction goal.

11.3.2 Doing It Right

Here are seven basic guidelines for establishing an effective incentive/reward program to motivate the occurrence of safety-related behaviors and improve OSH:

1. The behaviors required to achieve a safety reward should be specified and perceived as achievable by all participants.
2. Everyone who meets the behavioral criteria should be rewarded.
3. It is better for many participants to receive small rewards than for one person to receive a big reward.

FIGURE 11.7 Raffle drawings that result in few "lucky" winners and many "unlucky" losers can do more harm than good.

4. The rewards should be displayed and represent safety-related achievement. Coffee mugs, hats, shirts, sweaters, blankets, or jackets with a safety message are preferable to rewards that will be hidden, used, or spent.
5. Contests should not reward one group at the expense of another group.
6. Groups should not be penalized or lose their rewards for failure by an individual.
7. Progress toward achieving a safety-related reward should be systematically monitored and publicly posted for all participants.

Guideline 2 recommends against the popular lottery or raffle drawing. As illustrated in Figure 11.7, a lottery results in one "lucky" winner being selected and a large number of "unlucky" losers. The announcement of a raffle drawing might get many people excited, and if lottery tickets are dispensed for specific safe behaviors, there is some motivational benefit. Eventually, however, the valuable reward is received by a lucky few. Also, there is a disadvantage in linking chance with safety. It is bad enough we use the word "accident" in the context of OSH, as explained earlier in Chapter 3.

I have worked with a number of safety directors who used a lottery incentive/reward program and vowed they would never do it again. The big raffle prize, such as a snowmobile, pick-up truck, or television set, was displayed in a prominent location. Everyone got excited—temporarily—about the possibility of winning. Their attention was directed, however, at the big prize instead of the real purpose of the program: to keep everyone safe. *The material reward in an incentive program should not be perceived as the major payoff.* Incentives are only reminders to do the right thing, and rewards serve as feedback and a statement of appreciation for doing the right thing.

More important than external rewards is the way they are delivered. Rewards should not be perceived as a means of controlling behavior but as a declaration of sincere gratitude for making a contribution. If many people receive this recognition, you have many deposits in the emotional bank accounts of potential actively caring participants. That is why it is better to reward many than a few (Guideline 3).

Intervening with Consequences 129

FIGURE 11.8 Rewards with safety messages are special to those who earn them.

When rewards include a safety logo or message (Guideline 4), they become activators for OSH when displayed as illustrated in Figure 11.8. Also, if the OSH message or logo was designed by representatives from the target population, the reward takes on special meaning (as discussed previously in Chapter 10). Special items like these cannot be purchased anywhere and—from the perspective of internal consequences—they are more valuable than money.

As portrayed in Figure 11.9, contests that pit one group against another can lead to an undesirable win/lose situation (Guideline 5). Safety needs to be perceived as win/win. This means

FIGURE 11.9 Safety contests can motivate unhealthy competition.

developing a contract of sorts between every employee that makes everyone a stockholder in achieving a TSC. Everyone in the organization is on the same team. Team performance within departments or work groups can be motivated by providing team rewards or bonuses for team achievement. Every team that meets the "bonus" criteria should be eligible for the reward. In other words, Guideline 2 should be applied when developing incentive/reward programs to motivate team performance.

Obviously, developing and administering an effective incentive/reward program for OSH requires a lot of dedicated effort. There is no quick fix, but it is worth doing, if you take the time to do it right. As Aubrey Daniels (2000) wisely stated, "If you think this is easy, you are doing it wrong" (p. 179). Let's examine an exemplary case study.

11.3.3 An Exemplary Incentive/Reward Program

Many years ago, I consulted with the safety steering committee of a Hoechst Celanese company of about 2000 employees to develop a plant-wide safety incentive/reward program that followed each of the seven guidelines given previously. The steering committee, including four hourly and four salary employees, met several times to identify specific behavior-consequence contingencies. That is, they needed to decide what behaviors should earn what rewards. Their plan was essentially a "credit economy" where certain safe behaviors that could be performed by all employees earned a certain number of "credits."

At the end of the year, participants exchanged their credits for a choice of different prizes, all containing a special safety logo. The variety of behaviors earning credits included attending monthly safety meetings; participating in safety meetings; leading a safety meeting; writing, reviewing, and revising a job safety analysis; and conducting periodic OSH audits of environmental and equipment conditions and certain work practices. For a work group to receive credits for audit activities, the results of their environmental and PPE observations had to be posted in the relevant work areas. Only one behavior was penalized by a loss of credits—the late reporting of an injury.

At the start of the new year, each participant received a "safety credit card" for tallying ongoing credit earnings. Some individual behaviors earned credits for the person's entire work group, thus promoting group cohesion and teamwork. The audit aspect of this incentive/reward program exemplified a basic behavior-based principle for OSH—observation and feedback. Employees were systematically observed, and they received soon, certain, and positive feedback (a reward) after performing a target behavior.

Such an incentive/reward intervention is only one of several effective methods for increasing the occurrence of safe work practices with observation and feedback. In the next chapter, feedback is addressed more specifically as an external and extra consequence to prevent injuries.

11.3.4 Safety Thank-You Cards

I would be remiss if I did not describe "Safety Thank-You Cards" in a discussion of exemplary incentive/reward interventions. Over the years, I have seen a variety of thank-you cards designed by work teams and used successfully at a number of industrial sites, including Abbott Laboratories, Exxon Chemical, Ford, General Motors, Hercules, Hoechst Celanese, Kal Kan, Logan Aluminum, Phillip Morris, Westinghouse Hanford Company, and Weyerhaeuser.

At some locations, thank-you cards were used in a raffle drawing, exchangeable for food, drinks, or trinkets, or simply displayed on a plant bulletin board as a "Safety Honor Roll." Sometimes the cards could be accumulated and exchanged for tee shirts, caps, or jackets with messages or logos signifying safety achievement. At several plants, the person who delivered a thank-you card returned a receipt naming the recognized employee and describing the behavior

Intervening with Consequences

that earned the consequence, thus creating objective information to define a "Safe Employee of the Month."

At a few locations, the thank-you cards took on a special actively caring meaning. Specifically, when deposited in a special collection container, each thank-you card was worth 25¢ toward corporate contributions to a local charity or to needy families in the community. The actively caring card used at the Hoechst Celanese plant in Rock Hill, South Carolina, is shown in Figure 11.10. The back of the card included a colorful peel-off symbol that the recognized employee could affix in any number of places as a personal reminder of the recognition. I was surprised but pleased to see a large number of those thank-you stickers on employees' hard hats. Clearly, this actively caring thank-you approach to safety recognition has great potential as an inexpensive but powerful process to motivate occurrences of safe behavior.

Motivational leverage with this simple actively caring thank-you card was illustrated many years ago at the Hercules chemical plant in Portland, Oregon. The actively caring cards delivered for safety-related behaviors were similar to the one illustrated in Figure 11.10, except the peel-off sticker depicted the company logo—a rhinoceros. The incentive/reward contingency was simply stated. Give an Actively Caring Thank-You Card and "Rhino Sticker" to anyone who goes beyond the call of duty for safety or health.

Here is special motivational leverage. Every actively caring card received and then deposited in a designated "Actively Caring for Others" box was worth $1.00 to purchase toys for disadvantaged children in and around Portland. With this program, the 64 line workers at this chemical plant contributed more than $1750 during the Christmas holidays of 1996. Guess who picked out and delivered the toys? Children of the employees. Now that is special actively-caring-for-safety leverage from a simple behavior-based incentive/reward intervention.

FIGURE 11.10 This Actively Caring Thank-You Card offers reward leverage (see text).

11.4 IN SUMMARY

Writing this book and two follow-up editions was challenging, tedious, overwhelming, tiresome, sacrificing, and exhausting. Observers were apt to say I was self-directed and *intrinsically* motivated. Of course, I know better, and so do you.

Incidentally, I literally hand-wrote the various drafts of this text. I have never learned to type, and therefore, I have never benefited from the technological magic of computer word-processing. My colleagues explain that it is not necessary to be a skilled typist to reap the many *intrinsic* benefits of preparing a manuscript on a computer. "I type slowly with only one finger," some would say, "and I still enjoy the wonderful benefits of high-tech computer word-processing. I could never go back to preparing a manuscript by hand. You don't know what you're missing."

I am sure you have noticed my disparate uses of "intrinsic" in the two prior paragraphs, and you now understand the two meanings of this popular motivational term. Are my friends and colleagues so enthusiastic about computer-based word-processing because of intrinsic (internal) motivation or because of intrinsic (natural) consequences linked to their computer use? I hope you agree with the second reason. But, of course, the natural rewarding consequences can lead to positive self-talk and internal (unobservable) motivation.

Word-processing on a computer allows for rapid "quick-fix" control of letters, words, sentences, and paragraphs. Computer users also can walk to a printer and obtain a typed, "hard copy" of their document for study, revision, or dissemination. All of these soon, certain, positive consequences are connected naturally to word-processing behavior. No wonder my friends and colleagues are motivated about computer word-processing and have urged me to get on the high-tech "band wagon."

Actually, for this third edition, I read my hand-written additions to voicemail on my cell phone, and subsequently sent the email to my computer for processing. Thus, this senior citizen has made some intrinsically reinforcing progress with digital communication. Plus, the automatic corrections of spelling and spacing errors—along with checking "synonyms" for a better choice of words—have provided invaluable intrinsic reinforcement throughout the preparation of this third edition. Of course, that is only the beginning of using generative artificial intelligence (AI) or ChatGPT to make written expression more efficient and effective.

12 Intervening as an Actively Caring Coach

This chapter presents the principles and procedure of safety coaching—a key behavioral science process for OSH improvement. The letters of COACH represent the critical sequential steps of actively caring coaching: Care, Observe, Analyze, Communicate, Help. This coaching process is clearly relevant for improving behaviors in domains other than safety and in settings other than the workplace. Indeed, behavior-based feedback is critical for improvement in everything we do. This chapter shows you how to deliver feedback effectively.

> There are risks and costs to a program of action. But they are far less than the long-range risks and costs of comfortable inaction.
>
> **(John F. Kennedy)**

Large-scale behavior change is impossible without actively caring intervention agents—people willing and able to step in on behalf of another person's safety. In a TSC everyone feels responsible for OSH, pursuing it daily. They go beyond the call of duty to identify at-risk conditions and behaviors, and they intervene to correct them. This chapter is about becoming an actively caring intervention agent for OSH.

In simple terms, this means observing and supporting safe behavior and observing and correcting at-risk behavior. It might involve designing and implementing a particular intervention process for a work team, organizational culture, or an entire community. Or it might mean merely engaging in behavior-focused communication between an observer (the intervention agent) and the person observed. This is actively caring coaching, and to be effective, certain principles and guidelines need to be followed.

12.1 INTERVENING AS A SAFETY COACH

Coaching is essentially a process of one-on-one behavioral observation and feedback. The coach systematically observes the behaviors of another person and provides behavioral feedback on the basis of the observations. Safety coaches recognize and support the safe behaviors they observe, and they offer corrective feedback to reduce the occurrence of any observed at-risk behavior. This chapter specifies the steps of actively caring coaching, points out trainable skills needed to accomplish the process effectively, and illustrates tools and support mechanisms for increasing effectiveness.

The term "coach" is very familiar to us in an athletic context. In fact, winning coaches practice the basic observation-and-feedback processes needed for effective OSH coaching. They follow most of the guidelines reviewed here. As illustrated in Figure 12.1, the most effective team coaches observe the ongoing behaviors of individual players and record their observations in systematic fashion, using a team roster, behavioral checklist, or videotape.

Football coaches, for instance, spend many hours analyzing film. Then, they deliver specific and constructive feedback to team members in order to instruct, support, or motivate occurrences of desirable behavior and/or to decrease occurrences of undesirable behavior. Sometimes the feedback is given in a group session, perhaps by critiquing videotapes of team competition. At other times, the feedback is given individually in a personal one-on-one conversation. Usually, the one-to-one format has greater impact on individual performance.

FIGURE 12.1 Systematic observation and feedback are key to effective coaching.

The five letters of the word COACH can be used to remember the basic components of actively caring coaching—whether coaching the members of a winning athletic team or the individuals in a work group striving for an injury-free workplace. This is my favorite instructional acronym, because it not only contains the components of an effective coaching process, it lists them in the sequence in which they should occur.

12.1.1 "C" for Care

Caring is the underlying motivation for interpersonal coaching. Safety coaches truly care about the health and safety of their coworkers and they act on such caring. In other words, they "actively care" for OSH. When people realize by the words and body language of a coach that s/he cares, they are more likely to listen to and accept the coach's advice. When people know you care, they care what you know.

Our emotional bank accounts. Stephen Covey (1989) explained the value of interdependence among people—exemplified by actively-caring-for-safety coaching—with the metaphor of an "emotional bank account." People develop an emotional bank account with others through personal interaction. Deposits are made when the holder of the account views a particular interaction to be positive, as when s/he feels recognized, appreciated, or listened to. Withdrawals from a person's emotional bank account occur whenever that individual feels criticized, humiliated, or less appreciated, usually as a result of personal interaction.

Sometimes, it seems necessary to state extreme displeasure with another person's behavior. However, if such negative discourse occurs on an "overdrawn or bankrupt account," such corrective feedback will have limited impact. In fact, continued withdrawals from an overdrawn account can lead to defensive or countercontrol reactions. The person will simply ignore the communication or actually do things to discredit the source or undermine the process or system implicated in the communication.

Thus, OSH coaches need to demonstrate a caring attitude throughout their personal interactions with others. This maintains a healthy emotional bank account—operating in the "black." The woman in Figure 12.2 is requesting a deposit along with the withdrawal. Our emotional reaction to police officers depends on the proportion of deposits vs. withdrawals we have experienced with them.

Intervening as an Actively Caring Coach   135

BUT OFFICER, DON'T I GET CREDIT FOR BUCKLING UP?

FIGURE 12.2 Our attitude toward police officers would be more positive if we received deposits along with withdrawals.

A shared responsibility. People are often unwilling to coach or to be coached for safety because they view safety from an individualistic perspective. To them, it is a matter of individual or personal responsibility. This is illustrated by the verbal expression or internal script, "If Molly and Mike want to put themselves at risk, that's their problem, not mine." For some people, a change in personal attitude or perspective is needed in order to motivate them to coach others for OSH. People need to consider safety coaching a shared responsibility to prevent injuries throughout an entire work culture. This requires a shift from an individualistic to a collectivistic mindset. However, this is not easy, as reflected in the insightful poem, "The Cookie Thief" by Valerie Cox, reproduced in Figure 12.3.

The Cookie Thief
By Valerie Cox

A woman was waiting at an airport one night,
With several long hours before her flight.
She hunted for a book in the airport shop,
Bought a bag of cookies and found a place to drop.

She was engrossed in her book, but happened to see,
That a man beside her, as bold as could be
Grabbed a cookie or two from the bag between,
Which she tried to ignore, to avoid a scene.

She read, munched cookies and watched the clock,
As the gutsy "cookie thief" diminished her stock.
She was getting more irritated as the minutes ticked by,
Thinking, "If I wasn't so nice, I'd blacken his eye!"

With each cookie she took, he took one too.
When only one was left, she wondered what he'd do.
With a smile on his face and a nervous laugh,
He took the last cookie and broke it in half.

He offered her half, as he ate the other.
She snatched it from him and thought, "Oh, brother,
This guy has some nerve and he's also rude,
Why, he didn't even show any gratitude!"

She had never known when she had been so galled
And sighed with relief when her flight was called.
She gathered her belongings and headed for the gate,
Refusing to look back at the "thieving ingrate."

She boarded the plane and sank in her seat,
Then sought her book, which was almost complete.
As she reached in her baggage, she gasped with surprise.
There was her bag of cookies in front of her eyes!

"If mine are here" she moaned with despair,
"Then the others were his and he tried to share!"
Too late to apologize, she realized with grief,
That she was the rude one, the ingrate, the thief!

FIGURE 12.3 Independence from one person can stifle interdependence from another.

Many people accept a collectivistic or team attitude when it comes to work productivity and product quality. Coaching for production or quality is part of the job, but coaching for personal safety is often perceived as meddling. People need to understand that safety-related behaviors require as much, if not more, interpersonal observation and feedback as any other job activity.

One way to convince people to accept and support safety coaching as a shared responsibility is to point out their organization's injury record for a certain period of time. While an injury did not happen to them, it did happen to someone, and everyone certainly cares about that. Given this underlying caring attitude, the challenge is to convince others that effective safety coaching by them will reduce injuries to their coworkers. This is enabled by a behavior-based accountability system, as discussed next.

12.1.2 "O" FOR OBSERVE

OSH coaches observe the behavior of others objectively and systematically, with an eye for supporting safe behavior and correcting at-risk behavior. Behavior that illustrates going beyond "the call of duty" for the safety of another person should be especially supported. This is the sort of behavior that contributes significantly to OSH improvement and can be increased through supportive feedback. As illustrated in Figure 12.4, an OSH observer does not hide or spy and always asks permission first. Only with permission should an observation process proceed.

Observing behavior for supportive and corrective feedback is easy if the coach:

1. Knows exactly what behaviors are desired and undesired (an obvious requirement for athletic coaching) and
2. Takes the time to observe occurrences of those behaviors in the work setting.

FIGURE 12.4 Safety coaches are up-front about their intentions and ask permission before observing.

It is often advantageous—and usually essential—to develop a checklist of safe and at-risk behaviors and to rank them in terms of risk. Ownership and commitment are increased when workers develop their own behavioral checklists.

Developing a critical behavior checklist. Observation checklists can be generic or job-specific. A generic checklist is used to observe behaviors that may occur during several jobs. A job-specific checklist is designed for one particular job. Deciding which items to include on a critical behavior checklist (CBC) is a very important part of the coaching process. A CBC enables coaches to look for critical behaviors. A critical behavior is a behavior that:

1. Has led to a large number of close calls or injuries in the past.
2. Could potentially contribute to a large number of close calls or injuries because many people perform that behavior.
3. Has previously led to a serious injury or a fatality.
4. Could lead to a serious injury or fatality.

If only a few behaviors are observed in the beginning—which is often a good way to start a large-scale coaching process—a CBC should be designed for only the most critical behaviors.

Several sources can be consulted to obtain behaviors for a CBC, including injury records, close-call reports, job-hazard analyses, standard operating procedures, rules and procedural manuals, and the workers themselves. People already know a lot about their own safety-related performance. They know which safety rules they sometimes ignore, and they know when a close call has occurred to themselves or to others because of at-risk behavior. In addition, it is often useful to obtain advice from the plant doctor, nurse, safety director, or anyone else who maintains injury statistics for the plant.

When starting out, do not develop an exhaustive checklist of critical behaviors. A list can get quite long in a hurry. A long list for one-on-one observations can appear overwhelming and can inhibit the process. As with anything that is new and needs voluntary support, it is useful to start small and build. With practice, people find a CBC easy to use, and they accept additions to the list. They will also contribute in valuable ways to refine the CBC—from clarifying behavioral definitions to recommending behavioral additions and substitutions. The development and use of a CBC reflect a continuous improvement process. Ongoing CBC development and refinement benefit coaching observations, and vice versa.

A work group on a mission to develop a CBC needs to meet periodically to select critical behaviors to observe. I have found the worksheet depicted in Figure 12.5 useful in beginning the development of a CBC. Through interactive discussions, work groups define safe and at-risk behaviors in their own work areas relevant to each category. The category for body positioning and protection, for example, includes specific ways workers should protect themselves from environmental or equipment hazards. This can range from using certain PPE to positioning one's body parts in certain ways to avoid possible injury.

Some categories in Figure 12.5 may be irrelevant for certain work groups, like locking or tagging out equipment or complying with certain permit policies. A work group might add another general procedural category to cover particular work behaviors. Notice that defining safe and at-risk behaviors results in safety training in the best sense of the word. Participants learn exactly what safe behaviors are needed for a particular work process.

A list of work behaviors covering all of the generic categories in Figure 12.5 can be extensive and overwhelming. That provides numerous opportunities for coaching feedback, but remember, it takes time and practice to observe behaviors reliably—and to feel comfortable when being observed while working. I have found it useful to start the observation procedure with a brief CBC of four or five behaviors and then build on this list with practice, group discussion, and more practice.

Scheduling observation sessions. There is no best way to arrange for coaching observations. The process needs to fit the setting and the work procedures. That can only happen if the workers

Operating Procedures	Safe Observation	At-Risk Observation
BODY POSITIONING/PROTECTING Positioning/protecting body parts (e.g., avoiding line of fire, using PPE, equipment guards, barricades, etc.).		
VISUAL FOCUSING Eyes and attention devoted to ongoing task(s).		
COMMUNICATING Verbal or nonverbal interaction that affects safety.		
PACING OF WORK Rate of ongoing work (e.g., spacing breaks appropriately, rushing).		
MOVING OBJECTS Body mechanics while lifting, pushing/pulling.		
COMPLYING WITH LOCKOUT/TAGOUT Following procedures for lockout/tagout		
COMPLYING WITH PERMITS Obtaining, then complying with permit(s). (e.g., confined space entry, hot work, excavation, open line, hot tap, etc.).		

FIGURE 12.5 A worksheet to develop a generic critical behavior checklist.

themselves decide on the frequency and duration of the observations and derive a method for scheduling the coaching sessions. I have seen the protocol for effective coaching observations vary widely across plants and across departments within the same plant. The success of those processes did not vary substantially as a function of the protocol.

The 350 employees at one Exxon Chemical plant, for example, designed a process that counted on people to schedule their own coaching sessions with any two other employees. On days and at times selected by the person to be observed, two observers showed up at the individual's worksite and used a CBC to conduct a systematic, 30-minute observation session. That plant started with only one scheduled observation session per month for each worker, and observers were selected from a list of volunteers who had received special coaching training. One year later, employees scheduled two observation sessions per month, and any plant employee could be called on to coach. With slight periodic revisions, that interpersonal coaching process was in place for eight years (1993–2001) and enabled this plant to reach and sustain a record-low injury rate.

The Exxon procedure was markedly different from the "planned 60-second actively caring review" implemented at a Hoechst Celanese facility. For that one-on-one coaching process, all employees attempted to complete a one-minute observation of another employee's work practices in five general categories: body position, personal apparel, housekeeping, tools/equipment, and operating procedures. The initial plant goal was for each of the 800 employees to complete one 60-second behavioral observation every day. Results were entered into a computer file for a behavioral safety analysis of the work culture.

The CBC used for those one-minute coaching observations is shown in Figure 12.6. The front of each card included the five behavioral categories, a column to check "safe" or "at-risk" per category, and columns (Feedback Targets) to write comments about the observations. Those comments facilitated a behavioral feedback session following the observations, if it was convenient.

Intervening as an Actively Caring Coach

Observer: _____		Location: _____		Date: _____	
Audit Category	Safe	At Risk	Feedback Targets: Safe		At Risk
Position					
Safe Apparel					
Housekeeeping					
Tools/Equip.					
Procedures					
Total					

Front of One-Minute Audit Card

Observation Targets	Safe	At Risk	Observation Targets	Safe	At -Risk
Position * Line of Fire * Falling * Pinch Points * Lifting			**Tools/Equip.** * Condition * Use * Guards		
Safe Apparel * Hair * Clothes * Jewelry * PPE			**Procedures** * SOPs * JSAs * Permits * Lockout * Barricade * Equipment Release		
Housekeeping * Floor * Equipment * Storage of Materials					

Back of One-Minute Audit Card

FIGURE 12.6 Employees used this critical behavior checklist for one-minute observations.

The back of this CBC included examples (memory joggers) related to each behavioral category on the front of the card. Those examples summarize the category definitions developed by the CBC steering committee and determine whether "safe" or "at-risk" should be checked on the front of the card.

Critical features of the observation process. Duration, frequency, and scheduling procedures of CBC observations vary widely. Still, there are a few common features. First and foremost, the observer must ask permission before beginning an observation process. The name of the person observed should not be recorded. To build trust and increase participation, a "No" following a request to observe must be honored.

Asking permission to observe serves notice to work safely and therefore biases the observation data, right? In other words, when workers give permission to be coached, their attention to OSH will likely increase and they will try to follow all safety procedures. However, it is possible for people to overlook safety precautions, even when trying their best. They could be unconsciously at risk. When people give permission to be coached, their willingness to accept and appreciate feedback is maximized, even when the feedback is corrective.

What if people were to sneak around and conduct behavioral observations with no warning? That would be an unbiased plant-wide audit of work practices. It might even be accepted if those workers observed were not identified. However, if one-on-one feedback coaching is added to this procedure, an atmosphere of mistrust can develop.

Safety coaching should not be a way to enforce rules or play "gotcha." It needs to be seen as a process to help people develop safe work routines through supportive and corrective feedback. Giving corrective feedback after "catching" an individual off-guard performing an at-risk behavior will likely lead to defensiveness and lack of appreciation, even for a well-intentioned effort. It can also reduce interpersonal trust and alienate a person toward the entire safety-coaching process.

Feedback is essential. Each observation process with a CBC provides for tallying and graphing results as "percent safe behavior" on a group feedback chart. All checks for safe observations can be added and divided by the total number of checks (safe plus at-risk behaviors). The result is multiplied by 100 to yield a "percent safe behavior" score.

Applying this formula to checks written on the front of the CBC shown in Figure 12.6 results in a conservative estimate of the percentage of safe behavior. That is because a safe check mark on the CBC in this application means that every separate behavior of a certain category was marked safe on the back. Thus, this calculation requires all behaviors relating to a particular observation category to be safe for a "safe" designation. A calculation based on individual behaviors rather than on an "all-or-none" classification of a "safe" or "at-risk" category generally results in higher percentages.

There is no best way to do these calculations. However, it is important for the participants to understand the meaning of the feedback percentages. As shown in the lower half of Figure 12.7, these percentages can be readily displayed on a group feedback chart or graph. While feedback percentages are informative, it is vital to realize that the process is more important than the numbers.

The true value of the coaching process is not in the behavioral data, which are likely biased by uncontrollable factors, but in the actively caring interactions between employees. I have actually seen observers get so involved recording the numbers, such as frequency of safe and at-risk behaviors, that they let coworkers continue to perform an at-risk behavior while they observe and check columns on a CBC.

FIGURE 12.7 Feedback from a critical behavior checklist can be given one on one and in groups.

Intervening as an Actively Caring Coach

An individual's safety must come before the numbers in any observation process. When observers see an at-risk behavior that immediately threatens a person's health or safety, they should intervene at once. They can usually pick up the observation process afterwards. On the other hand, if the CBC was partially completed before the observer steps in to stop an at-risk behavior, it might be most convenient to communicate the other observations, especially if there are some safe behaviors to report. This "deposit" will help compensate for the "withdrawal" that was probably implicated by the need to stop a risky behavior.

12.1.3 "A" for Analyze

When interpreting observations, safety coaches draw on their understanding of the ABC contingency—for activator–behavior–consequence—introduced Chapter 8, as well as the behavioral science principles discussed in Chapter 9. They realize observable reasons usually exist for why safe or at-risk behaviors occur. They know certain unsafe behaviors are influenced by activators such as work demands, risky example-setting by peers, and inconsistent messages from management. They also appreciate the fact that at-risk behaviors are often motivated by one or more natural consequences, including comfort, convenience, work breaks, and approval from peers or work supervisors.

Such understanding is critical if OSH coaching is to be a "fact-finding" rather than "fault-finding" process. It also leads to an objective and constructive analysis of the situations observed. This is how people discover the reasons behind at-risk behaviors and design interventions to decrease them.

An ABC analysis can be accomplished before giving feedback to the person observed or during the one-on-one feedback process. Discussing the activators and consequences that possibly influenced certain work practices can lead to environmental or system improvements for decreasing occurrences of at-risk behavior:

- Was the behavior observed activated by a work demand or a desire to go on break or leave work early?
- Does the design of the equipment or the environment, or the ergonomic design of the task, influence at-risk behavior?
- Is certain PPE uncomfortable or difficult to use?
- Are fellow workers or supervisors activating at-risk behavior by requesting or demanding an excessive work pace?
- Are certain individuals motivating at-risk behavior from others by giving rewarding consequences, like words of appreciation for work done quickly at the expense of safety?

The actively caring coach explores answers to these and other questions with the person observed in the next phase of safety coaching—the heart of the process.

12.1.4 "C" for Communicate

A good coach is a good communicator. This means being an active listener and persuasive speaker. Because none of us are born with these skills, communication training sessions that incorporate role-playing exercises can be invaluable in developing the confidence and competence needed to deliver and receive behavioral feedback effectively. Such training should emphasize the need to separate behavior (actions) from person-states (attitudes and feelings). This enables corrective feedback without affecting feelings.

People need to understand that anyone can be at-risk without even realizing it, as in "unconscious incompetence," and performance can only improve with behavior-specific feedback. Once this fact is established, corrective feedback that is delivered appropriately will be appreciated, regardless of who is giving the feedback. Work status is not a factor.

The right delivery. I remember key aspects of effective verbal presentation with the "SOFTEN" acronym listed in Figure 12.8. First, it is important for the observer to initiate communication with a friendly smile and an open (flexible) perspective. "Territory" reflects the need to respect the fact that you are encroaching upon another person's work area. You should ask the person where it would be appropriate and safe to talk. It is also important to maintain a proper physical distance during this interaction.

Standing too close or too far from another person can cause interference and discomfort. There are prominent cultural differences in interpersonal distance norms. In the U.S., communicating closer than 18 inches with another person—measured nose to nose—is considered an intimate distance, with 0 to 6 inches reserved for comforting, protesting, lovemaking, wrestling, and other full-contact behaviors. The far phase of the intimate zone (6 to 18 inches) is used by individuals who are on very close terms.

Safety coaches in the U.S. should most likely communicate at a "personal distance" (18 inches to 4 feet). According to research by Edward Hall (1966), the near phase of the personal distance (18 to 30 inches) is reserved for those who are familiar with one another and on good terms. The far phase of the personal zone (2.5 to 4 feet) is typically used for social interactions between friends and acquaintances. This is likely to be the most common interaction zone for a workplace coaching session. Some coaching communication might occur at the near phase (4 to 7 feet) of Hall's social distance, which is typical for unacquainted individuals interacting informally. These distance recommendations are not hard-and-fast rules of conduct but rather personal territory norms to consider.

The "E" of our acronym represents three important directives to remember when coaching—energy, enthusiasm, and empathy. Your energy and enthusiasm can activate concern and caring on the part of the person with whom you are communicating. We all know that excited, committed coaches can make "true believers" out of the troops and that indifferent or distracted coaches

> Smile
> Open
> Friendly
> Territory
> Energy
>   Enthusiasm
>   Empathy
> Name

FIGURE 12.8 Principles of effective sending can be remembered with SOFTEN.

Intervening as an Actively Caring Coach 143

FIGURE 12.9 Body language communicates more than words.

can have the opposite effect. Actually, as depicted in Figure 12.9, our body language can speak louder than words.

Perhaps the most critical E-word in Figure 12.8 is "empathy," which is reflected in this profound quotation from Stephen R. Covey (1989): "Seek first to understand, before being understood" (p. 235). Whether the topic is empathic coaching, empathic listening, empathic discipline, or empathic leadership, the focus is on the other person's perceptions, needs, and/or feelings. When we begin conversations with this mindset, we can customize our coaching to fit the other person's perspective and be most successful at getting our point across.

When observing another individual's behavior, it is critical to try and view the context and the circumstances from that person's perspective. When you listen to explanations or excuses for at-risk behavior, try to see yourself in the same situation. Imagine what defense mechanism(s) you might use to protect your ego or self-esteem. And, when considering an action plan for improvement, you should attempt to view various alternatives through the eyes of the other person.

Finally, we need to remember that the dearest word to anyone's ears is his or her own name. Refer to the other person by name, but make it clear that the behavioral observations you have recorded will remain anonymous.

Individual feedback. Whether the aim is to support or correct, feedback should be specific and timely. It should specify a particular behavior and occur soon after the target behavior is performed. Also, it should be private, given one-on-one to avoid any interference or embarrassment from others. Corrective feedback is most effective if the alternative safe behavior is specified and potential solutions for eliminating the at-risk behavior are discussed.

Anyone giving feedback must actively listen to reactions with empathy. This is how a safety coach shows sincere concern for the feelings and self-esteem of the person on the receiving end of feedback. The best listeners give empathic attention with facial cues and posture, paraphrase to check understanding, prompt for more details, accept stated feelings without interpretation, and avoid arrogance such as, "When I worked in your department, I always worked safely."

Figure 12.10 reviews the critical characteristics of effective supportive and corrective feedback. This figure can be used as a guide for group practice sessions. Because it is not easy to give

> **Use Rewarding Feedback to Support Safe Behavior**
>
> - Give the feedback one-on-one and privately.
> - Give the feedback as soon as possible after the observation process.
> - Identify the safe behavior(s) observed.
> - Be sincere and genuine.
> - Express personal appreciation for setting the right example for others.
>
> **Use Correcting Feedback to Decrease At-Risk Behavior**
>
> - Give the feedback one-on-one and privately.
> - Give the feedback as soon as possible after the observation process.
> - Begin with acknowledgment of safe behavior(s) observed.
> - Identify the at-risk behavior(s) observed.
> - Specify the safe alternative to the at-risk behavior(s).
> - Indicate concern for the person's welfare.
> - Request commitment to avoid the at-risk behavior(s).

FIGURE 12.10 Maximize the beneficial impact of rewarding and correcting feedback with these key points.

safety-related feedback properly and because many people feel awkward or uncomfortable doing it, practice is important.

As a valuable group exercise, I recommend asking individuals to role-play the delivery and the reception of supportive and corrective safety-related behavioral feedback. Subsequent to a role-play, ask the observers to give the presenters behavioral feedback—first supportive feedback and then corrective feedback, all in the spirit of continuous improvement.

12.1.5 "H" for Help

The word "help" summarizes what safety coaching is all about. The purpose is to help an individual prevent injury by supporting safe work practices and correcting at-risk practices. It is critical, of course, that a coach's help is accepted. The four letters of HELP provide strategies to increase the probability that a coach's advice, directions, and corrections will be appreciated.

Humor. Safety is certainly a serious matter, but sometimes a little humor can add spice to our communications, increasing interest and acceptance. It can take the sting out what some find to be an awkward situation. In fact, researchers have shown that laughter can reduce distress and even benefit our immune system.

Esteem. People who feel inadequate, unappreciated, or unimportant are not as likely to go beyond "the call of duty" to benefit the safety and/or the health of themselves or others as are people who feel capable and valuable (See Chapter 15 for support of this premise.). The most effective coaches choose their words carefully, emphasizing the positive over the negative, in order to build or avoid lessening another person's self-esteem. Although Figure 12.11 is humorous, it is unfortunately an accurate portrayal of the atmosphere in many organizational cultures, including the university environment in which I have worked for more than 50 years.

Listen. One of the most powerful and convenient ways to build self-esteem is to listen attentively to another person with empathy. This sends the signal that the listener cares about the

Intervening as an Actively Caring Coach   145

FIGURE 12.11   Standard feedback more often depreciates than appreciates a person's self-esteem.

person and his or her situation. Plus, it builds self-esteem—"I must be valuable to the organization because my opinion is considered." After a safety coach listens actively, his or her message is more likely to be heard and accepted.

Praise. Praising others for their specific accomplishments is another powerful way to build self-esteem. If the praise targets a particular behavior, the probability of the behavior recurring increases. This reflects the basic principle of positive reinforcement and motivates people to continue their safe work practices and look out for the safety of their coworkers. Behavior-focused praising is an influential rewarding consequence, which not only increases the behavior it follows but also increases a person's self-esteem. Such supportive feedback, in turn, increases the individual's willingness to actively care for the safety of others, as is explained more completely in Part 5 of this text.

Human nature directs more attention to mistakes than successes. Errors stick out and disrupt the flow, so they attract reaction and attempts to correct them. However, as illustrated in Figure 12.12, when things are going smoothly—and safely—there is usually no stimulus to signal success. A person's good performance is typically taken for granted. We need to resist the tendency to go with the flow and sometimes express sincere appreciation and gratitude for ongoing safe behavior. The next chapter provides specifics on how to do that effectively.

12.2   IN SUMMARY

Safety coaching is a key intervention process for cultivating, achieving, and sustaining a TSC. In fact, the more employees who effectively apply the principles of actively caring behavioral coaching discussed here, the closer an organization will come toward achieving a TSC. The same is true for preventing injury in the community and among our immediate family members at home. Indeed, we need to practice the principles of safety coaching in every situation where an injury could occur following at-risk behavior.

Systematic OSH coaching throughout a work culture is certainly feasible in most settings. Large-scale success requires time and resources to develop materials, train necessary personnel,

FIGURE 12.12 People need frequent supportive feedback.

establish support mechanisms, monitor progress, and continually improve the process and support mechanisms whenever possible. For example, at the start of developing an initial action plan for safety coaching, the following questions need to be answered:

- Who will develop the critical behavior checklist (CBC)?
- How extensive will the first CBC be?
- What information will be used to define critical behaviors?
- How will safety coaches be trained with practice and behavioral feedback?
- How many coaches will be trained initially, and how can additional people volunteer to participate as a safety coach?
- How will the coaching sessions be scheduled, how often will people be coached, and how long will the coaching sessions last?
- Where will the group feedback graphs be posted, and who will be responsible for preparing the displays of safe-behavior percentages?
- Who will be on the steering committee to: oversee the safety coaching process, answer these and other questions about process implementation, maintain records, monitor progress, and refine procedural components whenever necessary?

This does not cover all of the issues, yet the list might appear overwhelming at first. There is no formula for a quick-fix solution. Organizational cultures vary widely according to personnel, history, policy, the work process, environmental factors, and current contingencies. So, implementation procedures need to be customized. Continuous input is necessary from the people protected by the coaching process—those whose long-term participation is needed.

It is likely to take significant time, effort, and resources to achieve a company-wide safety coaching process. With this end in mind, I recommend starting small to build confidence and optimism on small-win accomplishments. Then, with patience and diligence, set long-term goals for continuous improvement. Remember to celebrate achievements that reflect successive approximations of your vision—an actively caring organization of people who consistently coach each other effectively to increase occurrences of safe work practices and decrease occurrences of at-risk behaviors.

13 Intervening with Supportive Conversation

Interpersonal conversation defines the culture in which we work. It can create conflict and build barriers to OSH improvement or it can cultivate the kind of work culture needed to make a major breakthrough in injury prevention. Interpersonal conversation also affects our intrapersonal conversations or self-talk, which in turn influences our willingness to get involved in an OSH-improvement process. This chapter explains the reciprocal impact of interpersonal and intrapersonal conversation and offers guidelines for aligning both toward the achievement of a Total Safety Culture.

> Leadership is the ability to persuade others to do what you want them to do because they want to do it.
>
> **(Dwight Eisenhower)**

This chapter is about interpersonal conversation and coaching, but unlike Chapter 12 on systematic safety coaching, the emphasis here is on brief informal communication to support safe behavior and help such desirable behavior become more fluent. How we talk with others (interpersonal communication) influences their attitude and ongoing behavior, and how we talk to ourselves (intrapersonal communication) influences our own behavior and attitude. Therefore, this chapter also addresses self-talk—the mental scripts we carry around in our heads before, during, and after our behavior.

It is fair to say that the nature of our safety-related conversations with others can influence their degree of involvement in OSH. The variety of the safety-related conversations we have with ourselves influences whether we feel accountable to someone else for our safe behavior or whether we feel self-accountable for our safety-related behavior. This is the distinction introduced in Chapter 9 between feeling accountable or other-directed vs. feeling responsible or self-directed.

Bottomline: The long-term success of any effort to prevent injuries in the workplace, in the home, and on the road is determined by conversation.

13.1 THE POWER OF CONVERSATION

The dramatic success companies have experienced with behavior-based safety is essentially due to an increase in the quality and quantity of safety correspondence—not the high-tech communication referred to in Figure 13.1 but one-to-one interpersonal conversation about OSH. Such improvement, in turn, benefits people's self-talk about OSH, increasing their sense of personal control and optimism regarding their ability to prevent occupational injuries.

This chapter offers guidelines and techniques for getting more beneficial impact from our conversations with others and within ourselves. Four types of safety management are presented, each defined by the nature of interpersonal conversation.

13.2 THE ART OF IMPROVING CONVERSATION

Interpersonal conversation is a powerful tool that shapes personal and team attitudes about loyalty, commitment, social support, and OSH. Each of the techniques presented here can get

FIGURE 13.1 The power of conversation comes from face-to-face communication.

employees more involved in OSH and improve the overall level of safety-related performance in an organization.

Applying these techniques can also improve how you talk to yourself—your "self-talk." The payoff is increased self-esteem and perceptions of empowerment, which are essential for increasing our willingness to actively care for the safety and health of ourselves and others. Consider how much our self-esteem is influenced by how we talk to ourselves (intrapersonal) about how we think others talk about us (interpersonal).

13.2.1 Do Not Look Back

Has this ever happened to you? You ask for more safety involvement from a particular individual and you get a reaction like, "I offered a safety suggestion three years ago and no one responded." Have you ever attended a safety meeting where people spent more time reviewing past accomplishments or failures than discussing future possibilities and deriving action plans?

Those are examples of conversations stuck in the past. The discussion might be enjoyable, but little or no progress is made. Conversations about past events do help us connect with others and recognize similar experiences, opinions, and motives, but such communication does not enable progress toward problem-solving or continuous improvement. For that to happen, the conversation must leave the past and move on.

To direct the flow of a conversation from the past to future possibilities and then to the development of a practical action plan for the present, you should first recognize and appreciate what the other person has to say about the past. Then shift the focus toward the future. Remember, you are approaching this person to discuss possibilities for safety improvement and specific ways to get started now.

13.2.2 Seek Commitment

You know your interpersonal conversation is especially productive when someone makes a commitment to improve in a certain way. This reflects success in moving conversation from the past

Intervening with Supportive Conversation

to the future, and then to the present for a specific action plan. A verbal commitment also tells you that something is happening on an intrapersonal level within that other person. That person is becoming self-motivated, increasing the probability the target behavior will occur.

Now you can proceed to talk about how that commitment can be supported or how to hold the individual accountable. For example, one person might offer to help a coworker meet an obligation through verbal reminders, or an individual might agree to honor a commitment by showing a safety coach behavioral records that indicate improvement. That is, of course, the kind of follow-up conversation that facilitates personal achievement.

13.2.3 STOP AND LISTEN

In their eagerness to prevent an injury, safety advocates often give corrective feedback in a top-down, controlling manner. In other words, passion for safety sometimes leads to an overly directive approach to get others to change their behavior. You know from personal experience that a less directive approach to giving advice is often more effective, especially over the long term.

Think about it. How do you respond when someone overtly tells you what to do? Now it certainly depends on who is giving the directions, but I bet your reaction is not entirely positive. You might follow those instructions, especially if they come from someone with the power to control your consequences, but how will you feel? Will you be motivated to make a permanent change? You might if you asked for the direction, but if you did not request feedback, you could feel insulted or embarrassed.

Corrective feedback interpreted as an "adult-child" confrontation will probably not work. The supervisor in Figure 13.2 means well, but the worker does not perceive it that way. When a directive conversation is interpreted as controlling or demeaning, it is essentially ineffective. So, play it safe. Try to be more nondirective when using interpersonal conversation to affect behavior change.

FIGURE 13.2 Corrective feedback can feel demeaning.

13.2.4 ASK QUESTIONS FIRST

Instead of telling people what to do, try this. Get them to tell you, in their own words, what they ought to be doing in order to be safe. You could do this by asking questions with a sincere and caring demeanor. Avoid at all costs a sarcastic or demeaning tone, but first, point out certain safe behaviors you noticed—it is important to emphasize the positives.

Then move on to the seemingly at-risk behavior by asking, "Is there a safer way to perform that task?" Of course, you hope for more than a "yes" or "no" response to a question like that. However, if that is all you get, you need to be more precise in follow-up questioning. You might, for example, point out a particular work routine that seems risky and ask whether there is a safer way.

I recommend approaching a corrective feedback conversation as if you do not know the safest operating procedures, even though you think you do. You might, in fact, find your presumptions to be imperfect. The "expert" on the job might know something you do not know. If you approach the situation with that mindset, you will not get the kind of reaction given by the woman in Figure 13.3.

By asking questions, you are always going to learn something. If nothing else, you will hear the rationale behind taking a risk over choosing the safer alternative. You might uncover a barrier to OSH that you can help the person overcome. A conversation that entertains ways to remove obstacles that hinder safe behavior is especially valuable if it translates possibilities into feasible and relevant action plans.

You will know your nondirective approach to correction worked if your colleague owns up to his or her mistake, even under a cloud of excuses. Remember, it is only natural to offer a rationale for taking a risk. It is a way of protecting self-esteem. Let it pass, and remind yourself that when someone admits a mistake before you point it out, there is a greater chance for both acceptance of responsibility and improved behavior.

13.2.5 TRANSITION FROM NONDIRECTIVE TO DIRECTIVE

What if the person does not give a satisfactory answer to your questions about safer alternatives? What if the individual does not seem to know the safest operating procedure? Now you need to shift the conversation from nondirective to directive. You need to give behavior-focused advice.

FIGURE 13.3 Diagnosis requires questions and answers.

Intervening with Supportive Conversation 151

In this case, start with a phrase like, "As you know," as my friend John Drebinger (2000) advises. Open the conversation with a phrase that implies the person really does know the safe way to perform, but for some reason s/he just overlooked it (or forgot) this time. That could happen to anyone. Such an opening can help prevent others from feeling their intelligence or safety knowledge has been insulted.

13.2.6 BEWARE OF BIAS

Every conversation you have with someone is biased by prejudice or prejudgment filters—in yourself and within the other person. You cannot get around it. From personal experience, people develop opinions and attitudes and these, in turn, influence subsequent experience. With regard to interpersonal conversation, we have subjective prejudgment filters that influence what words we hear, how we interpret those words, and what we say in response to those words. In Chapter 5, this bias was introduced as "premature cognitive commitment." Every conversation influences how we process and interpret the next conversation.

Figure 13.4 illustrates such bias. The female driver is merely trying to inform the other driver of an obstacle in the road, but that is not what the driver of the pick-up hears. That driver's prior driving experiences led to a biased interpretation of the warning. You could call such selective listening an "autobiographical bias" (Covey, 1989). Of course, factors besides prior experience can bias interpersonal communication, including personality traits, mood states, physiological needs, and future expectations.

It is probably impossible to escape completely the impact of this premature bias in our conversations, but we can exert some control. Actually, each of the conversation strategies discussed here is helpful. For example, the nondirective approach attempts to overcome this bias by listening with empathy and asking relevant questions before giving instruction. With this approach, a person's biasing filters can be identified and considered in the customization of a plan for corrective action.

Pay close attention to the body language and verbal tone in these interpersonal conversations. As you have heard before, the method of delivery can hold as much or more information as the words themselves. Listen for passion, commitment, or caring. If nothing else, you could learn

FIGURE 13.4 Selective listening can be hazardous to your health.

whether the messenger understands and believes the message, and you may learn a new way to deliver a message yourself. Bottomline: Our intrapersonal conversations can either facilitate or hinder what we learn from our interpersonal conversations.

13.3 RECOGNIZING DESIRABLE BEHAVIOR

The "Flow of Behavior Change" model presented in Chapter 9 (see Figure 9.6) indicates that supportive intervention through interpersonal recognition or supportive feedback is critical for making safe behavior more fluent. Thus, supportive safety conversations are needed throughout a work culture, since most of us need to be more fluent at performing some safety-related behaviors. Yet, this type of conversation is relatively rare.

We are more likely to use our interpersonal conversations to correct rather than support or recognize. In fact, we are more inclined to beat ourselves up for our own mistakes, instead of celebrating our personal successes. Now, how can this be explained? Why do we pay more attention to the negative things in our lives?

One reason is that the mistakes stick out. They upset the flow and are readily noticed. When people are doing the right thing, the process runs smoothly, and we keep on going. We go with the flow. We hardly notice the variety of good behaviors occurring at the time. Instead of being a "good finder," we wait for an observable mistake and then make our move.

Another reason for a focus on the negative is that some people believe they and others learn best from their mistakes. They think paying attention to errors is the best way to improve performance. Behavioral research tells us this is wrong. We learn best when we get supportive feedback or recognition for doing the right thing. As discussed in Chapter 11, positive consequences are good for our attitude. You know how you feel when you get recognition—when it is genuine. You feel good, and that is what we need. We need people feeling good about themselves when they go out of their way for OSH.

We need to have the same mindset about safety that the gold prospectors had about their challenge. Their focus was on finding gold. They sifted to find the good, not the bad. Likewise, we need to prospect for good behavior, even when bad behavior might be more obvious. Mom has the right idea in Figure 13.5.

FIGURE 13.5 Prospect for the good in others.

Intervening with Supportive Conversation

After finding good behavior, it is important to give supportive recognition appropriately. Most of us have not been taught how to give recognition effectively. Our commonsense is not sufficient. Behavioral research, however, has revealed strategies for making interpersonal recognition most rewarding. When you know how to maximize the impact of your recognition, you might use this powerful communication intervention more often. Listed in Figure 13.6 are seven guidelines for giving quality recognition. Let's consider each one in order.

> **How to Give Supportive Recognition**
> ❏ Recognize during or immediately after safe behavior.
> ❏ Make it personal for both parties.
> ❏ Connect specific behavior with general higher-level praise.
> ❏ Deliver it privately and one-on-one.
> ❏ Let it stand alone and soak in.
> ❏ Use tangibles for symbolic value only.
> ❏ Second-hand recognition has special advantages.

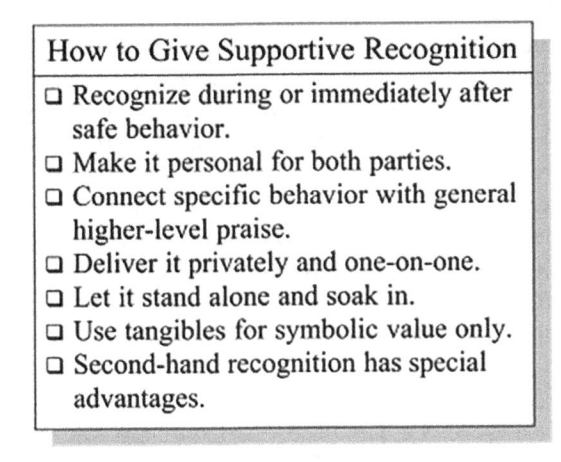

FIGURE 13.6 Follow seven conversation guidelines when giving recognition to support safety achievement.

13.3.1 RECOGNIZE DURING OR IMMEDIATELY AFTER DESIRABLE BEHAVIOR

In order for recognition to provide optimal direction and support, it needs to be associated directly with the desired behavior. People need to know what they did to earn the appreciation. If it is necessary to delay the recognition, then the conversation should relive the activity that deserves recognition. Reliving the behavior means talking specifically about what behavior warrants the positive attention. You could ask the person you are recognizing to describe aspects of the situation and the desirable behavior. This facilitates direction and motivation to continue the behavior. When you connect a person's behavior with recognition, you also make the supportive conversation special and personal.

13.3.2 MAKE RECOGNITION PERSONAL FOR BOTH PARTIES

A supportive conversation is most meaningful when it is personal. The recognition should not be general appreciation that could fit anyone in any situation. Rather, it needs to be customized to fit a particular individual and circumstance. This happens naturally when the recognition is linked to a specific behavior.

When you recognize someone, you are expressing personal thanks. It is tempting to say "we appreciate" rather than "I appreciate" and to refer to company gratitude instead of personal gratitude. Speaking for the company can come across as impersonal and insincere. Of course, it is appropriate to reflect value to the organization when giving praise, but the focus should be personal. "I saw what you did to support our safety process and I really appreciate it. Your example illustrates actively caring and demonstrates the kind of leadership we need around here to achieve a Total Safety Culture." This second statement illustrates the next guideline for giving quality recognition.

13.3.3 CONNECT SPECIFIC BEHAVIOR WITH GENERAL HIGHER-LEVEL PRAISE

A supportive conversation is most memorable when it reflects a higher-order characteristic. Adding a quality attribute like leadership, integrity, trustworthiness, or actively caring to the recognition statement makes the recognition more rewarding and most likely to increase the kind of intrapersonal communication that boosts self-esteem. It is important to state the specific behavior first and then make a clear connection between that behavior and the positive attribute it reflects.

13.3.4 Deliver Recognition Privately and One-On-One

Because effective recognition is personal and indicative of higher-order attributes, it needs to be delivered in private. After all, the recognition is special and only relevant to one person. So, it will mean more and seem more genuine if it is given from one individual to another.

It seems conventional to recognize individuals in front of a group. This approach is typified in athletic contests and reflected in the pop psychology slogan, "Praise publicly and reprimand privately." Many managers take the lead from this commonsense statement and give their individual recognition at group meetings. Is it not maximally rewarding to be held up as an exemplar in front of one's peers? Not necessarily—many people feel embarrassed when receiving special attention in a group setting. Plus, if positive recognition is rare in a particular culture, receiving public recognition in front of one's peers can feel awkward and fuel a win/lose mindset among the participants.

13.3.5 Let Recognition Stand Alone and Soak In

Keep a supportive conversation simple and to the point. Give your behavior-based praise a chance to soak in. In this fast-paced era of trying to do more with less, we try to communicate as much as possible when we finally get in touch with a busy person. After recognizing a person's special safety effort, we are tempted to tag on a bunch of unrelated statements, even a request for additional behavior. This comes across as, "I appreciate what you've done for safety, but I need more." Resist the temptation to do more than praise the good behavior you saw with sincere gratitude. If you have additional points to discuss, it is better to reconnect later, after your praise has had a chance to sink in and become a part of the individual's self-talk or intrapersonal conversation.

13.3.6 Use Tangibles for Symbolic Value Only

Tangibles can detract from the self-motivation aspect of quality recognition. If the focus of a recognition process is placed on a material reward, the words of appreciation can seem less significant. In turn, the impact on one's intrapersonal conversation is lessened.

On the other hand, tangibles can add to the quality of interpersonal recognition if they are delivered as tokens of appreciation. As discussed in Chapter 11, if tangibles include a safety slogan, they can help to promote OSH. But how you deliver a trinket will determine whether it adds to or subtracts from the value of your supportive conversation. The benefit of your praise is weakened if the tangible is viewed as a payoff for the safety-related behavior. On the other hand, if the tangible is seen as symbolic of going beyond the call of duty to actively care for safety, it strengthens the praise.

13.3.7 Special Advantages of Secondhand Recognition

Up to this point, the situation is one person recognizing another person directly for a particular safety-related behavior. It is also possible to recognize a person's outstanding efforts indirectly, and such an approach can have special benefits. Suppose, for example, you overhear a colleague talking to another person about your outstanding safety presentation. How will such secondhand recognition affect you? Will you believe the words of praise were genuine?

Intervening with Supportive Conversation

Sometimes people are suspicious of the genuineness of praise when it is delivered face-to-face. The receiver of praise might feel, for example, there is an ulterior motive to the recognition. The deliverer of praise might be expecting a favor in return for the special recognition. Perhaps both individuals had recently attended the same behavior-based safety course, and the verbal exchange is viewed as only an extension of a communication exercise. In these cases, one-on-one recognition can be devalued as sincere appreciation. Secondhand recognition, however, is not as easily tainted with such potential biases. Therefore, its genuineness is less suspect.

My point here is that gossip can be good—if it is *positive*. When we talk about the success of others in behavior-specific terms, we begin a cycle of positive communication that can influence the continuation of desired behavior. It also helps to build an internal script for self-motivation. We also set an example for the kind of interpersonal and intrapersonal conversations that increase self-esteem, empowerment, and group cohesion. As explained in Part 5 of this book, these are the very person-states that increase occurrences of actively caring for safety behavior and cultivate the achievement of a TSC.

13.4 RECEIVING RECOGNITION WELL

The list of guidelines in Figure 13.6 for giving quality recognition is not exhaustive, but it does cover the basics. Following those guidelines will increase the benefit of a conversation to support desirable performance. The most important point is that more recognition and gratitude for OSH-related behavior is needed in every organization, whether given firsthand or indirectly through positive gossip. It only takes a few seconds to deliver quality behavioral recognition.

Most of us get so little positive recognition from others that we are caught completely off guard when acknowledged for our actions. We do not know how to accept appreciation when it finally comes. Some claim they do not deserve the special recognition. Others actually accuse the person giving recognition of being insincere or wanting something from them. This can be quite embarrassing to the person doing the recognizing and could discourage that person from giving more recognition.

Remember the basic motivational principle that consequences influence the behaviors they follow. Well, this is true for both the person giving recognition and the person receiving recognition. Quality recognition increases occurrences of the behavior being recognized, and one's reaction to the recognition influences whether the recognizing behavior is likely to reoccur. Thus, it is crucial to react appropriately when we receive recognition from others. Seven basic guidelines for receiving recognition are listed in Figure 13.7. Here is an explanation for each of these.

How to Receive Supportive Recognition

- Avoid denial and disclaimer statements.
- Listen actively with genuine appreciation.
- Relive recognition later for self-motivation.
- Show sincere appreciation.
- Recognize the person for recognizing you.
- Embrace the reciprocity principle.
- Ask for recognition when it is deserved but not forthcoming.

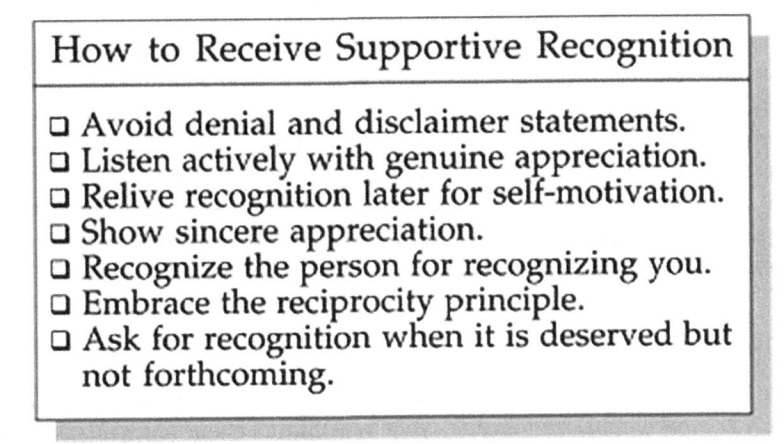

FIGURE 13.7 Follow seven conversation guidelines when receiving recognition in order to increase the occurrence of interpersonal support.

13.4.1 Avoid Denial and Disclaimer Statements

Whenever I attempt to give quality recognition, whether to a colleague, student, waiter/waitress, or hotel clerk, the most common reaction I get is awkward denial. Some act as if they did not hear me and keep doing whatever they are doing, or they offer a disclaimer like, "It really was nothing special," "Just doing my job," or "Other members of our team deserve more credit than I."

These days, the most common response I hear following a sincere expression of gratitude is "No problem." What does that common, seemingly habitual retort to a "Thank you" imply, and what are the ramifications? Yes, that reply seems to mean the "Thank you" was completely unnecessary and perhaps unappreciated. Such a response is certainly not a positive reinforcer that will increase the probability of subsequent expressions of gratitude.

Plus, it certainly does not promote reciprocity or a "pay it forward" response from the recipient. How about following a statement of gratitude with the simple "You're welcome" response our parents taught us, with "You would do the same for me." That rejoinder promotes an actively caring mindset rather than stifling "pay it forward" reciprocity.

We need to accept recognition without denial and disclaimer statements, and we should not deflect the credit to others. It is okay to show pride in our small-win accomplishments, even if others contributed to the successful outcome. After all, the vision of a TSC includes everyone going beyond the call of duty for their own safety and for the safety of others. In this context, most people deserve recognition on a daily basis. It is not "employee of the month," it is "employee of the moment"—moment-by-moment.

13.4.2 Listen Attentively with Genuine Appreciation

Listen proactively to the person giving you recognition. You want to know what you did right. Plus, you can evaluate whether the recognition is given well. If the recognition does not pinpoint a particular behavior, you might ask the person, "What did I do to deserve this?" This will help to improve that person's method of giving recognition.

Of course, it is important to not appear critical, but rather to show genuine appreciation for the special attention. Consider how difficult it is for most people to go out of their way to recognize others. Then, revel in the fact that you are receiving some recognition, even if its quality could be improved. Remember that the person who recognizes you is showing gratitude for what you did, and s/he will appreciate you more if you accept the recognition well.

13.4.3 Relive Recognition Later for Self-Motivation

Obviously, most of your safety-related behaviors go unnoticed. You perform many of those behaviors when no one else is around to observe you. And even when other people are available, they will likely be so preoccupied with their own routines that they will not notice your extra effort. So, when you finally do receive some recognition, take it in as well-deserved. Remember the many times you have gone the extra mile for OSH but did not get noticed.

You need to listen intently to every word of praise, not only to show you care but also because you want to remember this special occasion. Relive this moment later by talking to yourself. Such self-recognition can motivate you to continue going beyond the call of duty for OSH.

13.4.4 Show Sincere Appreciation

After listening actively with humble acceptance, you need to show sincere gratitude with a smile and a "Thank you." As emphasized earlier, your reaction to being recognized can determine whether similar recognition will occur again. So be prepared to say something to reflect pleasure in the special conversation. I find it natural to add "You've made my day," to a "Thank you"

Intervening with Supportive Conversation

because it is the truth. When people go out of their way to offer me quality recognition, they have made my day, and I often relive such situations to improve a later day.

13.4.5 RECOGNIZE THE PERSON FOR RECOGNIZING YOU

When you accept recognition well, you reward the person giving support for his/her extra effort. This can motivate that individual to do more recognizing. Sometimes, you can do even more to increase quality recognition. Specifically, you can recognize the person for recognizing you. In this case, you apply quality recognition principles to reward certain aspects of the supportive conversation. You might say, for example, that you really appreciate the pinpointing of a certain behavior and the reference to higher-order praise. Such rewarding feedback provides direction and motivation for those aspects of the recognition process that are especially worthwhile and need to become routine.

13.4.6 EMBRACE THE RECIPROCITY PRINCIPLE

Some people resist receiving recognition because they do not want to feel obligated to give recognition to others. That is the reciprocity norm at work. If we want to cultivate a TSC, we need to embrace this norm. Research has shown that when you are nice to others—as when providing them with special praise—you increase the likelihood they will reciprocate by showing similar behavior. You might not receive the returned favor, but someone will. They will "pay it forward."

Bottomline: Realizing your genuine acceptance of quality recognition will activate the reciprocity norm, and the more this norm is activated from positive interpersonal conversation, the greater the frequency of interpersonal recognition. So, accept recognition well and embrace the reciprocity norm. The result will be more interpersonal actively caring behavior consistent with the vision of a TSC.

13.4.7 ASK FOR RECOGNITION WHEN DESERVED BUT NOT FORTHCOMING

There is one final strategy I recommend for increasing occurrences of recognition conversations throughout a culture. If you feel you deserve recognition, why not ask for it? This might result in recognition viewed as less genuine than if it were spontaneous, but the outcome from such a request can be quite beneficial. You might receive some words worth reliving later for self-motivation. But most importantly, you will remind the other individual in a nice way that s/he missed a prime opportunity to offer quality recognition. This could be a valuable learning experience for that individual.

Consider the possible benefit from your statement to another person that you are pleased with a certain result from your extra effort. With the right tone and affect, such verbal behavior will not seem like bragging but rather a declaration of personal pride in a small-win accomplishment. The other individual will probably support your personal praise with relevant personal testimony, and that will boost your own self-motivation. Plus, you will teach the other person how to support the OSH-related behavior of others.

13.5 IN SUMMARY

I hope you are convinced that the status of OSH in your organization is greatly determined by how safety is talked about—from the managers' boardroom to the workers' breakroom. Whether we feel responsible for OSH and are committed to go for a breakthrough depends on our interpretation or mental script about safety-related conversations.

We often focus our interpersonal and intrapersonal conversations on the past. This helps us connect with others, but it also feeds our prejudice filters and limits the potential for conversation to facilitate beneficial change. We enable progress when we move our conversations with ourselves and with others from past happenings to future possibilities, and then to the development of a current action plan.

Expect people to protect their self-esteem with excuses for their past mistakes. Listen proactively for barriers to safe behavior reflected in such excuses. Then help the conversation shift to a discussion of possibilities for improvement and personal commitment to apply a feasible action plan. That is often more likely to occur with a nondirective than a directive approach whereby more questions are asked than directives given. It is also useful to use opening words to protect the listener's self-esteem and limit the impact of reactive bias.

Remember that everyone needs genuine recognition and appreciation now and then for doing the right thing, especially for OSH-related behavior. Following the guidelines given here for delivering and receiving quality recognition will increase the frequency of such supportive conversation and its interpersonal and intrapersonal benefits. William James, the first renowned American psychologist, wrote "the deepest principle in human nature is the craving to be appreciated" (from Carnegie, 1936, p.19).

Part Five

Actively Caring for People

14 Understanding Actively Caring

Actively caring is planned and purposeful behavior, directed at environment, person, or behavior factors. It is reactive or proactive, and it is direct or indirect. Direct, proactive, and behavior-focused actively caring is most challenging, but it is usually most important for large-scale injury prevention. This chapter discusses conditions and situations that inhibit actively caring behavior. We need to understand why people resist opportunities to actively care for safety. Then, we can develop interventions to increase occurrences of that desirable behavior—critical for achieving a Total Safety Culture.

> We cannot live only for ourselves. A thousand fibers connect us with our fellow men; and among those fibers, as sympathetic threads, our actions run as causes; and they come back to us as effects.
>
> **(Herman Melville)**

That quotation from Herman Melville appeared in a popular paperback entitled *Random Acts of Kindness* (page 31). There, the Editors of Conari Press (1993) introduced the idea of randomly showing kindness or generosity toward others for no ulterior motive except to benefit humanity. That notion seems quite analogous to the actively caring concept discussed earlier in various contexts. Indeed, a recurring theme in this book is that a TSC can only be achieved if people intervene regularly to protect and promote the safety and health of others. Actively caring, however, is usually not random. It is planned and purposeful; plus, as implied in Melville's assertion, actively caring behaviors (actions) are supported by positive consequences (effects).

Part 5 of this book addresses the need to increase actively-caring-for-people (AC4P) behavior throughout a culture and explains how to optimize the safety and health benefits from interventions designed and implemented to cultivate a TSC. Psychologists have identified conditions and individual characteristics (or person-states) that influence people's willingness to actively care for the safety and/or the health of others. These are revealed in this chapter and linked to practical things you can do to increase occurrences of AC4P behavior.

14.1 WHAT IS ACTIVELY CARING?

Figure 14.1 presents a simple flow chart summarizing the basic approach to culture change. We start a culture-change mission with a vision or ultimate purpose—for example, to achieve a TSC. With group consensus supporting the vision, we develop procedures or action plans to accomplish our mission. These are reflected in process-oriented goals established to activate goal-related behaviors.

Appropriate goal-setting (as detailed in Chapter 10), self-affirmations, and a positive attitude can activate behaviors to achieve goals and visions, but we must consider one of B. F. Skinner's most important legacies—"selection by consequences" (1981). As depicted in Figure 14.1, consequences are needed to support the right behaviors and correct wrong behaviors. Without support for the "right stuff," good intentions and initial efforts fade away. Sometimes, natural or intrinsic consequences are available to motivate desired behaviors, but often—especially for OSH—extrinsic consequences need to be managed to motivate occurrences of the behaviors required to achieve our goals.

However, vision, goals, and consequences are not sufficient for culture change. People need to actively care about relevant goals, action plans, and consequences. They need to believe in and own the vision. They need to feel obligated to work toward attaining goals that support the

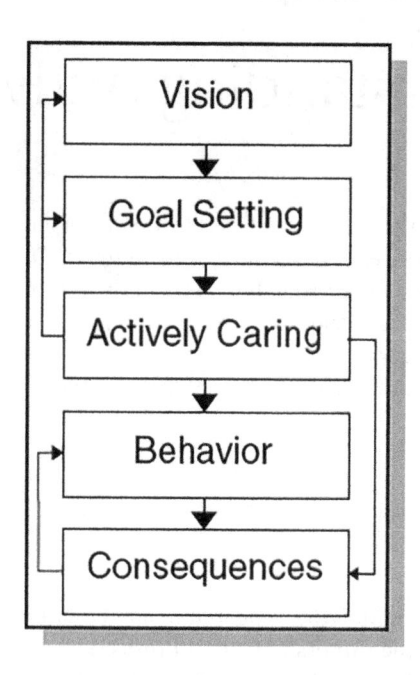

FIGURE 14.1 A Total Safety Culture requires vision and behavior management.

vision, and they need to give supportive and corrective feedback to increase the occurrence of behaviors consistent with vision-relevant goals. This is key to continuous improvement and to achieving a TSC.

14.1.1 Three Ways to Actively Care

The "Safety Triad" introduced in Chapter 3 is useful for categorizing AC4P behaviors. AC4P behavior can address environmental factors, person-states, or behavior. When people alter environmental conditions or reorganize or redistribute resources in an attempt to benefit others, they are actively caring from an environment perspective. Behaviors in this category include attending to housekeeping details, posting a warning sign near an environmental hazard, designing a guard for a machine, locking out the energy source to production equipment, and cleaning up a spill.

Person-based actively caring occurs when we attempt to make other people feel better. We address their emotions, attitudes, or mood states. Examples include proactively listening to others with empathy, inquiring with concern about another person's difficulties, complimenting an individual's personal appearance, and sending a "Get well" or "Thank you" card. This kind of actively caring is likely to boost a person's self-esteem, optimism, or sense of belonging—which, in turn, increases that individual's propensity to actively care for safety. This type of AC4P behavior is detailed in Chapter 15, which includes reactive behaviors performed in crisis situations. For example, if you pull someone out of an equipment pinch point or administer cardiopulmonary resuscitation, you are actively caring from a person-based perspective.

From a proactive perspective, behavior-based actively caring is most constructive and most challenging. This happens when you apply an instructive, supportive, or motivational intervention to improve another person's OSH-related behavior. The one-on-one coaching process described in Chapter 12 represents behavior-focused AC4P. Giving someone behavior-based recognition in a supportive conversation, as discussed in Chapter 13, is also actively caring with a behavior focus.

Understanding Actively Caring

14.1.2 Why Categorize AC4P Behaviors?

So, why go to the trouble of categorizing AC4P behaviors? Good question! I think it is useful to consider what these behaviors are trying to accomplish and to realize the relative difficulty in performing each of them. Environment-focused caring might be easiest for some people because it usually does not involve interpersonal interaction. When people contribute to a charity, donate blood, or complete an organ donor card, they do not interact personally with the recipient of the contribution. These behaviors are certainly commendable and may represent significant commitment and effort, but the absence of a personal encounter between giver and receiver warrants consideration separate from other types of AC4P behavior.

Certain conditions and personality traits might facilitate or inhibit one type of AC4P behavior and not the other. For example, communication skills are needed to actively care on the personal or behavioral level. Behavior-focused AC4P is more direct and usually more intrusive than person-focused AC4P. It is riskier and potentially confrontational to attempt to direct or motivate another person's behavior than it is to demonstrate concern, respect, or empathy for someone.

Classifying AC4P behaviors also provides insight into their benefits and liabilities. Both Brown (1991) and the editors of Conari Press (1993) recommend we feed expired parking meters to keep people from paying excessive fines. Let's consider the behavioral impact of this environment-focused "random act of kindness." What will the vehicle owner think when finding an unexpired parking meter? Could this lead to a belief that parking meters are unreliable and further one's mismanagement of time? Is there a price to pay in people becoming less responsible about sharing public parking spaces?

When considering the long-term and large-scale impact of some AC4P strategies, other approaches might come to mind. In the parking-meter situation, for example, the potential impact would be improved by adding some behavior-focused AC4P. Along with feeding the expired parking meter, why not place a note under the vehicle's windshield wiper explaining the act of kindness? The note might also include some time-management hints. This additional step might not only improve behavior but set an AC4P example. The recipient of the note would probably be more inclined to actively care for someone else.

Each type of AC4P behavior can be direct or indirect, with direct behavior requiring effective communication strategies. For instance, leaving a note to explain an AC4P act does not involve interpersonal conversation. Similarly, you can report an individual's safe or at-risk behavior to a supervisor and eliminate the need for one-to-one communication skills.

Similarly, person-focused AC4P behavior does not always involve interpersonal dialogue. You can send someone a get-well card or leave a friendly uplifting statement on an answering machine or by e-mail. It is also possible that environment-focused acts will include interpersonal interaction, as when delivering a contribution to a needy individual. This direct vs. indirect categorization of AC4P behavior is illustrated in Figure 14.2.

14.1.3 An Illustrative Anecdote

Several years ago, I was driving on a toll road in Norfolk, VA, on route to the Fort Eustis Army Base, where a transportation safety conference was being held. Three of my undergraduate research students were with me. Each was scheduled to give a 15-minute talk at the conference. This was to be their first professional presentation and they appeared quite nervous. They were paging frantically through their notes, making last-minute adjustments.

"Were you this nervous Doc, when you gave your first professional address?" one student asked.

"No, I don't think so," I replied in jest, "I was obviously better prepared."

"Can we just read our paper?" asked another.

"Absolutely not," I retorted. "Anyone could read your paper. It is much more professional and instructive to just talk about your paper informally with the audience."

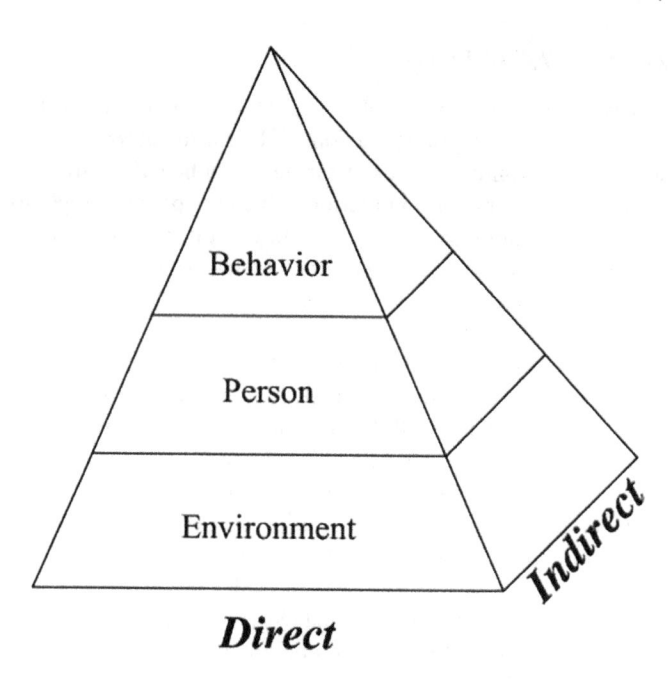

(Pyramid diagram with three tiers labeled, from top: Behavior, Person, Environment. The left face is labeled "Direct" and the right face is labeled "Indirect".)

FIGURE 14.2 Actively caring is usually most challenging and useful when direct and behavior focused.

Naturally, this conversation just caused more anxiety and distress for my students. Something had to be done to distract them—to break the tension. As we approached the first of several tollbooths along the highway, I thought of an AC4P solution. After paying a quarter for my vehicle, I handed the attendant another quarter and said, "This is for the vehicle behind us; the driver is using her safety belt and deserves the recognition." My students put down their papers and watched the attendant explain to the driver that I paid her toll because she was buckled up. Because I slowed down to observe this, the driver caught up with us, pulled next to us in the right lane, and acknowledged our AC4P behavior with a "shoulder belt salute"—a smile and tug on her shoulder strap.

At the next toll booth, the driver of the vehicle behind us was not buckled up, but that did not stop me. I gave the attendant an extra quarter and said, "This is for the vehicle behind us; please ask the driver to buckle up." Again, I slowed down to watch and, to our delight, the driver buckled up on the spot. When the vehicle passed us, the driver gave us a smile and a "thumbs up" sign.

I kept doing this at every tollbooth until exiting the highway, by which time my students had almost forgotten about their papers. They seemed relaxed and at ease when entering the conference room and each gave an excellent presentation. Later, we discussed how the toll-booth intervention took their minds off their papers and their anxiety.

Brown (1991) recommended that his son occasionally pay the toll for the vehicle behind him. This is redistribution of resources. It is also actively caring with an environment focus. By adding a safety-belt message, I was able to accomplish more than the "random act of kindness" suggested by Brown (1991) and the editors of Conari Press (1993). I was able to support those who were already buckled up and to influence some other drivers to buckle up. In other words, realizing the special value of behavior-based AC4P enabled me to get more benefit from an environment-focused strategy with very little extra effort. That behavior-based effort was particularly convenient and effortless because it was indirect. Note how the system for categorizing AC4P behavior enables us to compare real and potential acts of kindness and then to consider ways to increase their impact.

14.1.4 A Hierarchy of Needs

The hierarchy of needs proposed by the humanist Abraham Maslow (1943) is probably the most popular theory of human motivation. It is taught in a variety of college courses, including introductory classes in psychology, sociology, economics, marketing, human factors, and systems management. It is considered a stage theory. Categories of needs are arranged hierarchically, and we do not attempt to satisfy needs at one stage or level until the needs at the lower stages are satisfied.

First, we are motivated to fulfill our physiological needs, which include basic survival requirements for food, water, shelter, and sleep. After these needs are under control, we are motivated by safety and security needs—the desire to feel secure and protected from future dangers. When we prepare for future physiological needs, we are proactively working to satisfy our need for safety and security.

The next motivational stage includes our social acceptance needs—the need to have friends and to feel like we belong. When these needs are gratified, our concern focuses on self-esteem, the desire to develop self-respect, gain the approval of others, and achieve personal success.

When I ask audiences to tell me the highest level of Maslow's Hierarchy of Needs, several people usually shout "self-actualization." However, when I ask for the meaning of "self-actualization," I receive limited or no reaction. This is probably because the concept of being self-actualized is rather vague and ambiguous. In general terms, we reach a level of self-actualization when we believe we have become the best we can be, taking the fullest advantage of our potential as human beings. We are working to reach this level when we strive to be as productive and creative as possible. Once accomplished, we possess a feeling of brotherhood and affection for all human beings and a desire to help humanity as members of a single family—the human race. Perhaps, it is fair to say that those individuals are ready to actively care.

Maslow's Hierarchy of Needs is illustrated in Figure 14.3, but self-actualization is not at the top. Maslow (1971) revised his renowned hierarchy shortly before his death in 1970 and placed self-transcendence above self-actualization. Transcending the self means going beyond self-interest and is quite analogous to the AC4P concept. According to Viktor Frankl (1962), for

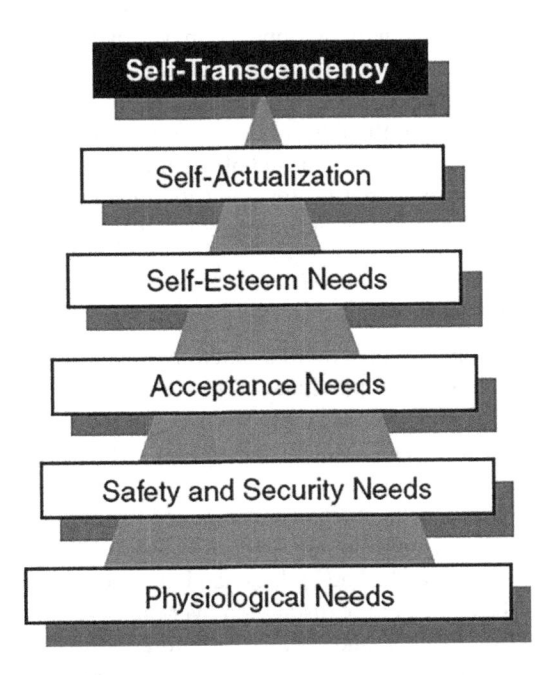

FIGURE 14.3 The highest need in Maslow's revised hierarchy reflects actively caring.

example, self-transcendence includes giving ourselves to a cause or another person and is the ultimate state of existence for the healthy person.

Similarly, the AI language model ChatGPT defines self-transcendence as

> the process or state of going beyond one's ordinary self, limitations, or narrow perspectives. It involves transcending the ego and expanding one's awareness, identity, and consciousness to embrace broader aspects of existence, such as spiritual or transcendent experiences, connection with others, or a greater understanding of the world.

Thus, after satisfying needs for self-preservation, safety and security, acceptance, self-esteem, and self-actualization, people can be motivated to reach the ultimate state of self-transcendence by reaching out to help others—to actively care. Safety leaders reach the top of Maslow's Hierarchy regularly—whenever they intervene to keep another person safe from personal injury.

While Figure 14.3 shows Maslow's Hierarchy of Needs with the correct level at the top, it is noteworthy that Maslow never represented his theory of motivation as a pyramid with a presumed lockstep progression of one need requiring satisfaction before a higher need becomes motivating. Instead, he considered human motivation an integrated hierarchy, whereby lower-level deficiency motives and higher-level growth motives coexist interdependently—a harmonious integration of the various need levels (cf. Kaufman, 2020).

Thus, people can be motivated to actively care for the safety of others before their own biological needs (e.g., for food, shelter, and sleep) are completely satisfied. Many readers have likely been in such situations. You have transcended your self needs in order to help another person, and as a result your sense of belonging or social acceptance need was enhanced, as well as your self-esteem.

14.2 THE PSYCHOLOGY OF ACTIVELY CARING

Walking home on March 13, 1964, Catherine (Kitty) Genovese reached her apartment in Queens, NY, at 3:30 a.m. Suddenly, a man approached her with a knife, stabbed her repeatedly and then raped her. When Kitty screamed, "Oh my God, he stabbed me! Please help me!" into the early morning stillness, lights went on and windows opened in nearby buildings. Seeing the lights, the attacker fled, but when he saw no one come to the victim's aid, he returned to stab Kitty Genovese eight more times and rape her again.

The murder and rape lasted more than 30 minutes and was witnessed by 38 neighbors. One couple pulled up chairs to their window and turned off their lights so they could get a better view. Only after the murderer and rapist departed for good did anyone phone the police. When the neighbors were questioned about their lack of intervention, they could not explain it.

14.2.1 Lessons from Research

Professors Bibb Latané and John Darley (1970) studied bystander apathy by staging emergency events observed by varying numbers of individuals. Then, they systematically recorded the speed at which one or more persons came to the victim's rescue. In the most controlled experiments, the participants sat in separate cubicles (as depicted in Figure 14.4) and could not be influenced by the body language of other participants.

In the first study of this type, the participants introduced themselves and discussed problems associated with living in an urban environment. In each condition, the first individual introduced himself and then casually mentioned he had epilepsy and that the pressures of city life made him prone to seizures. During the course of the discussion over the intercom, he became increasingly loud and incoherent, choking, gasping, and crying out before lapsing into silence. The researchers measured how quickly the subjects left their cubicles to help him.

Understanding Actively Caring

FIGURE 14.4 Participants in the Latané and Darley experiment could not see each other and thought they were conversing with one, two, or five other individuals.

When the participants believed they were the only witness, 85 percent left their cubicles within three minutes to intervene. However, only 62 percent of the participants who believed one other witness was present left their cubicle to intervene, and only 31 percent of those who thought five other witnesses were available attempted to intervene. Within three to six minutes after the seizure began, 100 percent of the lone participants, 81 percent of the participants with one presumed witness, and 62 percent of the participants with five other bystanders left their cubicles to intervene.

The reduced tendency of observers of an emergency to help a victim when they believe other potential helpers are available has been termed the *bystander effect* and has been replicated in several situations. Researchers have systematically explored reasons for the bystander effect and have identified conditions influencing this phenomenon. The results most relevant to OSH are reviewed here.

Diffusion of responsibility. It is likely many observers of the Kitty Genovese rape and murder assumed that another witness would call the police or attempt to scare away the assailant. Perhaps, some observers waited for a witness more capable than they to rescue Kitty. Does this factor contribute to lack of intervention for OSH? Do people overlook environmental hazards or at-risk behaviors in the workplace because they presume someone else will make the correction? Perhaps some people assume, "If the employees who work in that work area don't care enough to remove the hazard or correct the risk, why should I?"

A helping norm. Many, if not most, U.S. citizens are raised to be independent rather than interdependent. However, intervening for the benefit of others, whether reactively in a crisis situation or proactively to prevent a crisis, requires sincere belief and commitment toward interdependence. Social psychologists refer to a "social responsibility norm" as the belief that people should help those who need help. Research participants who scored high on a measure of this interdependence norm—as a result of nurturing during childhood or special training sessions—were more likely to intervene in a bystander intervention study, regardless of the number of other witnesses.

Knowing what to do. When people know what to do in a crisis, they do not fear making a fool of themselves and do not wait for another, more appropriate person to intervene. The bystander effect was eliminated when observers had certain competencies, such as training in first-aid treatment, which enabled them to take charge of the situation. In other words, when observers believed they had the appropriate tools to help, bystander apathy was decreased or eliminated.

This conclusion is also relevant for proactive or preventive action, as in safety-related intervention. When people receive tools to improve OSH, and believe those tools will be accepted and effective at preventing injuries, bystander apathy for OSH will decrease. This implies, of course, the need to promote a social responsibility or interdependence norm throughout a culture, and to teach and support specific intervention strategies or tools to prevent workplace injuries.

Interdependence. Researchers demonstrated reduced bystander apathy when observers knew one another and had developed a sense of belonging or mutual respect from prior interactions. Most, if not all, of the witnesses to Kitty Genovese's murder did not know her personally, and it is likely the neighbors did not feel a sense of comradeship or community with one another. Situations and interactions that reduce a we/they, or a territorial perspective, and increase feelings of belongingness or "community" will increase the likelihood of people looking out for each other—interdependent AC4P behavior.

Mood states. Several social psychology studies have found that people are more likely to offer help when they are in a good mood. The mood states that facilitated helping behavior were created very easily by arranging for potential helpers to: 1) find a dime in a phone booth, 2) receive a cookie, 3) watch a comedy film, or 4) experience pleasant aromas. How are these findings relevant for OSH?

Daily events can elevate or depress our moods. Some events are controllable, some are not. Clearly, the nature of our interactions with others can have a dramatic impact on the mood of everyone involved. Perhaps, remembering the research on mood and its effect will motivate us to adjust our interpersonal conversations with coworkers (see Chapter 13). We should also interact in a way that could influence a person's beliefs or expectations in certain directions, as explained next.

14.3 A CONSEQUENCE ANALYSIS OF AC4P BEHAVIOR

When I related the Kitty Genovese incident to my family and asked their opinions, I received a unanimous reaction that reflects the influence of consequences—a common theme throughout this book. My wife and two daughters proclaimed that most observers did not help that woman because they feared for their own safety. The perpetrator was armed with a knife. The onlookers could certainly see there was an emergency requiring specific assistance from anyone who would take responsibility.

According to an interpretation based on our understanding of the power of consequences, people resisted taking responsibility because they perceived that it could mean more trouble—or potential harm—than it was worth. It was safer to assume that someone else more capable would intervene. According to this consequence perspective, people hesitated to intervene because they perceived more potential costs than benefits, not because they were apathetic.

The matrix in Figure 14.5 combines two levels of cost (low vs. high) to the potential intervention agent and the victim in order to predict when AC4P behavior will occur. AC4P behavior is most likely (lower left cell of Figure 14.5) when the costs for helping are low, for example, convenient and not dangerous, and the costs to the victim for not helping are high, as when the victim is seriously injured. On the other hand, intervention is least likely when the perceived personal costs for intervening are high, for example, effortful and risky, and the apparent costs to the victim for no intervention are low, as when an experienced worker is performing at-risk behavior with no negative consequences.

Understanding Actively Caring 169

		Costs to Bystander for Intervening	
		Low	*High*
Costs to Victim for Not Intervening	*Low*	Depends on person factors	Intervention improbable
	High	Intervention probable	Diffusion of responsibility

FIGURE 14.5 Costs to bystanders for intervening and costs to a victim for not intervening determine the probability of intervention (adapted from Piliavin et al., 1981).

The Genovese incident fits the lower right cell of Figure 14.5—high perceived cost for both helper and victim. Although the costs for not helping Kitty were extremely high, resulting ultimately in her death, the costs for helping were also high, in fact, potentially fatal. This implies significant conflict for the person deciding what to do. The conflict can be resolved by helping indirectly—by telephoning police or an ambulance—or by reinterpreting the situation. For example, the observer could presume someone else will intervene—diffusion of responsibility—or perhaps rationalize that the victim does not deserve help.

A bystander might rationalize, for example, that Kitty Genovese should not have been walking the streets in that neighborhood at 3:30 in the morning. Rationalization reduces the perceived costs for not intervening and enables the bystander to ignore the situation without excessive shame or guilt. According to this cost-reward interpretation, when bystanders perceive high costs both for intervening and for not intervening in a crisis, they recognize the need for action, hesitate because of perceived personal costs, and then search for an excuse to do nothing, as depicted in Figure 14.6.

FIGURE 14.6 People give a variety of excuses for not helping.

The upper left quadrant of Figure 14.5 represents situations most analogous to actively caring for safety. Although a simple low-cost intervention might be called for to correct an environmental hazard or an at-risk behavior, there is no immediate emergency and, thus, no need for immediate action. There is low perceived cost if no action is taken: "We've been working under these conditions for months and no one has been hurt." Intervention in situations represented by this cost quadrant is most difficult to predict. Many factors can influence perceived consequences that are positive and negative, and small changes in these factors can tilt the cost-reward balance in favor of stepping in or standing back.

Through testimonials and constructive discussions, employees can be convinced that the potential cost of not intervening is higher than they initially thought. This can occur, for example, by considering the large degree of plant-wide exposure to a certain uncorrected hazard. Also, it might be worthwhile to remind people of the large-scale detrimental observational learning that could occur from the continuous performance of an at-risk behavior. Furthermore, education and role-playing exercises can reduce the perceived personal costs of AC4P behavior. It is also true that personal factors, such as mood states discussed earlier, can determine whether intervention occurs.

14.3.1 THE POWER OF CONTEXT

The influence of context in determining whether we actively care for another person's safety cannot be overemphasized. The context in which behavior occurs can affect one's evaluation of the costs and benefits of helping vs. not helping a victim. In other words, the perceived consequences of AC4P behavior depend to a significant degree on the environmental and social context in which the relevant behaviors occur. Let's look more closely at this context variable and consider its impact on OSH-related behavior.

According to my copy of *The American Heritage Dictionary* (1985), context refers to "the circumstances in which a particular event occurs" (page 316), including both the outside and inside stuff surrounding people when they are performing. This refers to what we see others doing on the outside and how we feel on the inside—from feelings of competence, confidence, and commitment to perceptions of insecurity, uncertainty, and risk.

Figure 14.7 is worth more than a thousand words to describe context. Have you seen a mild-mannered and polite person turn into an impatient and hostile creature after getting behind the wheel of an automobile? The environmental and competitive context of driving interacts with certain personality dispositions to produce "Mr. Hyde" on the road. Then, we have a nationwide epidemic of "road rage."

14.4 CONTEXT AT WORK

Does the mission statement of your industry reflect an overarching concern for quality and quantity of production? Is safety considered a priority (instead of a value) that gets shifted when production quotas are emphasized? Is safety viewed as a top-down condition of employment rather than an employee-driven process supported by management? Are OSH programs handed down to employees with directives to "implement per instructions" rather than "customize for your work area"?

Are safety initiatives discussed as short-term "flavor-of-the-month" programs rather than an ongoing process that needs to be continuously improved to remain evergreen? Are close-call and injury "investigations" perceived as fault-finding searches for a single "root cause" rather than fact-finding opportunities to learn what can be done to reduce the probability of personal injury? Are the components of an OSH initiative considered piecemeal factors independent of

FIGURE 14.7 In the context of driving, many individuals transition from mild-mannered to rude, hostile, and impolite.

other organizational functions rather than aspects of an organizational system of interdependent functions?

Are employees held accountable for outcome numbers that provide little direction for proactive change and personal control rather than process numbers that are diagnostic regarding the achievement of a TSC? Do employees take a dependency stance toward industrial safety whereby they depend on the organization to protect them with rules, regulations, engineering safeguards, and PPE?

A "Yes" answer to any of those questions implies contextual barriers that need to be overcome in order to achieve the ultimate TSC—an injury-free workplace. A "No" answer to all of those questions is symptomatic of a work context that encourages people to actively care for the health and safety of others. In such a work culture, it is not sufficient to rely on the organization's safe operating procedures or even on personal responsibility and self-discipline, but rather on interpersonal teamwork and a shared interdependent responsibility to protect each other. In such a work context, AC4P behavior can be cultivated and a TSC can be achieved.

14.5 IN SUMMARY

AC4P behavior is planned and purposeful. It can be direct or indirect and its focus is on the environment, person, or behavior. Actively caring that addresses the environment is usually easiest to perform because it does not require interpersonal communication and potential confrontation. Behavior-focused actively caring is often most proactive, but is most difficult to carry out effectively because it attempts to influence another person's behavior in a nonemergency situation.

The finding that people often refuse to actively care in a crisis, especially when they can share the responsibility of intervening with others, is quite analogous to most work settings. Hence, it is important to understand the factors that can influence this resistance, referred to as "bystander apathy." For example, people with a sense of social responsibility and comradeship for others at work, and who believe they have personal control in in an interdependent work culture, are more apt to intervene for the safety of others.

It is possible to increase those person characteristics among people through policy, procedures, and interpersonal communication. Enhancing those person-states and therefore one's willingness to actively care for OSH is key to achieving a TSC. Evidence-based practical techniques for making this happen are addressed in the next two chapters.

15 The Person-Based Approach to Actively Caring

Our willingness to actively care for others is affected by certain feelings and mindsets. If we have a strong sense of self-esteem, self-efficacy, personal control, optimism, and belonging, there is a greater chance we will go beyond the call of duty for OSH. Each of these person-states is explained in this chapter, and the research supporting direct relations between these person-states and AC4P behavior is reviewed. Understanding those connections enables us to design conditions and interventions to increase AC4P behavior throughout an organization or a community.

Our deeds determine us, as much as we determine our deeds.

(George Eliot)

That quotation by George Eliot indicates that our behaviors influence something about us and implies that good deeds or AC4P behaviors are good for us. They change something about us, and this, in turn, affects subsequent behavior. Does this mean our AC4P behaviors influence us to actively care even more? That is a nice thought and seems intuitive, but what does it really mean?

This chapter explores a host of questions arising from the human dynamics reflected in Eliot's words. What is it about us that changes as a result of our good deeds, and will that change lead to more good deeds? Can we influence people to actively care beyond just managing activators and consequences to directly influence behavior? In other words, can we change something about people that will inspire them to be more willing to actively care for the safety and health of others? If answers to these questions can be turned into practical procedures, we will know how to increase the frequency of AC4P behavior throughout a culture.

15.1 ACTIVELY CARING FROM THE INSIDE

Perhaps you recall earlier discussions in this text about "outside" vs. "inside" dynamics of people. Chapter 3, for example, distinguished between behaviors (outside) vs. intentions, attitudes, and values (inside), and emphasized that we should start with behavior. A prime principle of applied behavioral science (ABS) is that it is easier, especially for large-scale culture change, to "act a person into safe thinking" than it is to address attitudes and values directly in an attempt to "think a person into safe acting." Another key principle of ABS is that the consequences of our behavior influence how we feel about the behavior. Generally, positive consequences lead to good feelings or attitudes, whereas negative consequences lead to bad feelings or attitudes.

Long-term behavior change requires people to change inside as well as outside. The promise of a positive consequence or the threat of a negative consequence can maintain the desired behavior while the response-consequence contingencies are in place. What happens when the reinforcing consequences are withdrawn? What happens when people are in situations, like at home, when no one is holding them accountable for their behavior? If people do not *believe* in the safe way of doing something, and do not *accept* safety as a value or a personal mission, do not expect them to choose the safe way when they have the choice. In addition, if people are not self-motivated to keep themselves safe, do not expect them to actively care for the safety of others.

As discussed in Chapter 5, numerous internal and situational factors influence how we perceive activators and consequences. For example, if we see activators and consequences as nongenuine ploys to control us, our attitude about the situation will be negative. On the other hand, if we believe the external contingencies are genuine attempts to help us do the right thing, our attitude will be more positive. Thus, personal or internal dynamics determine how we receive activator and consequence information. This can influence whether environmental events facilitate or inhibit occurrences of safe behavior.

Chapters 10 and 11 revealed how direct manipulations of activators and consequences can influence behavior on a large scale. Now, let's consider how changes in internal person factors can benefit behavior. In particular, how do various inside factors or person-states influence actively caring for safety?

15.1.1 PERSON TRAITS VS. STATES

Some person factors are presumed to be *traits*, while others are *states*. Theoretically, traits are relatively permanent characteristics of people and they do not vary appreciably over time or across situations. In contrast, states are characteristics that can change moment-to-moment depending on circumstances and personal interactions. When our goals are thwarted, for example, we can be in a state of frustration. When experience leads us to believe we have little control over events around us, we can be in a state of apathy or helplessness. Those person-states can influence various behaviors.

Frustration often provokes aggressive behavior; perceptions of helplessness can inhibit constructive behavior or facilitate inactivity. In contrast, certain life experiences can affect positive person-states, such as optimism, personal control, self-confidence, and a sense of belonging. In turn, these person-states can facilitate occurrences of constructive behavior, including AC4P behavior.

15.1.2 AC4P STATES

AC4P dynamics internal to people are states, not traits. Plus, certain conditions—including activators and consequences—can influence these psychological states. These person-states are illustrated in Figure 15.1, a model my associates and I have used many times to stimulate

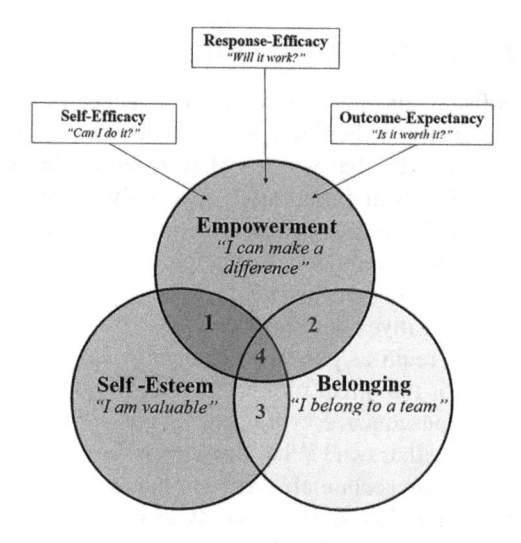

1. I can make a valuable difference.
2. We can make a difference.
3. I'm a valuable team member.
4. We can make valuable differences.

FIGURE 15.1 Certain person states influence an individual's willingness to actively care for the safety and health of others.

The Person-Based Approach to Actively Caring 175

one-to-one and group conversations among employees. We discuss specific situations, operations, or incidents that influence their willingness to actively care for the achievement of a TSC. Let's examine these influential person-states in more detail.

Self-esteem ("I am valuable"). How do you feel about yourself? Generally good or generally bad? Your level of self-esteem is determined by the extent to which you generally feel good about yourself. If we do not feel good about ourselves, it is unlikely we will care about making a positive difference in the lives of others. As illustrated in Figure 15.2, a person's self-esteem can get pretty low. The better we feel about ourselves, the more likely we will actively care for the safety and health of other people.

FIGURE 15.2 A person's self-esteem can get pretty low.

It is important to maintain a healthy level of self-esteem and to help others raise their self-esteem. Research shows that people with relatively high self-esteem report fewer negative emotions and less depression than do people with low self-esteem. Those with higher self-esteem also handle life's stresses better. Recall the discussion of stress vs. distress in Chapter 6. Higher self-esteem can turn stress into positive stress rather than negative distress.

Researchers have also found that individuals who score higher on measures of self-esteem are less susceptible to outside influences, are more confident of achieving personal goals, and they make more favorable impressions on others in social situations. Plus, in support of the AC4P model depicted in Figure 15.1, people with higher self-esteem help others more frequently than do those who score lower on a self-esteem scale (Batson et al., 1986).

Empowerment ("I can make a difference"). In the management literature, empowerment typically refers to delegating authority or responsibility, or sharing decision-making. In contrast, the person-based perspective of empowerment focuses on how the person who receives more power or influence reacts. From a psychological science perspective, empowerment is a matter of personal perception. Do you feel empowered or more responsible? Can you handle the additional assignment? This view of empowerment requires the belief that "I can make a difference."

A proper assessment of feeling empowered involves asking three questions derived from social learning theory (Bandura, 1982). The first question is, "Can I do it?" Do I have the knowledge, skills, ability, and resources to achieve a particular objective? A "Yes" answer to this first empowerment question reflects a personal belief in having the competence to make it happen. Albert Bandura (1982, 1997) called this "self-efficacy." Note that the term self-efficacy places the focus on *personal* belief. An observer might think an individual has the competence to complete a task, but the empowered individual might feel differently. Thus, a "Yes" answer to the first empowerment question implies a belief among those who receive an assignment that they have the relevant personal competence to complete the process and achieve the outcome goal.

Dozens of studies have found that participants who score relatively high on a measure of self-efficacy perform better at a wide range of tasks. They show more commitment to a goal and work harder to pursue it. They demonstrate greater ability and motivation to solve complex problems at work, and they are more apt to handle stressors positively rather than with negative distress.

Self-efficacy contributes to self-esteem, and vice versa; but these person-states are different. Simply put, self-esteem refers to a general sense of self-worth, whereas self-efficacy refers to feeling successful or effective at performing a particular task. Self-efficacy is more focused and can vary markedly from one task to another. One's level of self-esteem remains rather constant across situations.

The second question—"Will it work?"—reflects response-efficacy. Does the recipient of an assignment believe that performing the required behavior(s) will contribute to a valued mission for the performer and/or for others? In this case, education about the mission-based value of performing the task-relevant behavior may be needed. With regard to behavior-based safety, the response-efficacy question translates into believing an AC4P behavioral-coaching process will eventually contribute to preventing injuries, and help to cultivate a TSC.

Whereas a negative answer to the self-efficacy question indicates a need for more training, a negative answer to the response-efficacy question implies a need for education. People might believe they are able to accomplish a particular process or task (i.e., self-efficacy), but they might not believe such accomplishment will make a difference in a desired outcome (i.e., response-efficacy). In this case, they need education, including an explanation of an evidence-based principle or theory, and perhaps the presentation of convincing data. Regarding the AC4P-coaching process, substantial data are available to show that an AC4P behavior-observation-and-feedback process prevents workplace injuries effectively (e.g., Geller et al., 2004; Sulzer-Azaroff & Austin, 2000).

The third question—"Is it worth it?"—targets motivation. Will performing the task-relevant behavior(s) result in a worthwhile outcome—achieve a positive consequence and/or avoid a negative consequence? Obviously, motivation is crucial for an individual to feel empowered, and, as detailed in Chapter 11, appreciating the positive consequences to gain and/or the negative consequences to avoid motivates the occurrence of goal-relevant behavior.

Optimism. Although not included among the person-states depicted in Figure 15.1, this person-state also influences the quantity and the quality of AC4P behavior. Optimism is reflected in the statement, "I expect the best." It is the learned expectation that life events, including personal actions, will turn out well. Compared to pessimists, optimists maintain a sense of humor, perceive problems or challenges in a positive light, and plan for a successful future. They focus on what they can do rather than on how they feel. As a result, optimists handle stressors constructively and experience positive stress more often than negative distress. Optimists essentially expect to be successful at whatever they do, and so they work harder than pessimists to reach their goals. As a result, optimists are beneficiaries of the self-fulfilling prophecy. Figure 15.3 shows how an optimistic perspective can influence one's attempt to achieve more.

The self-fulfilling prophecy starts with a personal expectation about one's future performance and ends with that expectation coming true because the individual performs in such a way to make it happen. What do you expect when your boss or supervisor asks to see you? Do you expect the best? Our past experiences with top-down control and the use of negative consequences to

The Person-Based Approach to Actively Caring

FIGURE 15.3 Optimists expect more from their efforts.

influence our behavior often results in pessimistic rather than optimistic expectations. Moreover, our approach to this situation, illustrated in Figure 15.4, can support our negative expectations.

FIGURE 15.4 When our boss asks to see us, we expect the worst.

If you expect to be penalized or reprimanded every time your boss or supervisor calls you into the office, then your body language and demeanor will subtly reflect that expectation. You will "telegraph" those signals to your boss, who might think, "Scott sure looks guilty, I wonder what he's done that needs to be penalized?"

However, if you approach the interaction with an optimistic attitude, reflected in your body language and verbal behavior, the results could be more positive. You could, for example, write a different internal script. "No one is perfect, and I might have missed something. Everyone can improve with specific behavioral feedback. If I can help to make the interaction constructive, the outcome will be positive."

It is important to understand that fulfilling a pessimistic prophecy can depreciate one's feelings of empowerment, optimism, and self-esteem. With this realization, we should do whatever we can to make interpersonal conversations positive and constructive. This will not only increase optimism in a work culture, but will also promote a sense of group cohesiveness or a sense of belonging—another person-state that facilitates AC4P behavior.

Belonging. In his best seller, *The Different Drum: Community Making and Peace*, M. Scott Peck (1979) challenges us to experience a sense of true community with others. We need to develop feelings of belonging with one another regardless of our political preferences, cultural backgrounds, and religious doctrine. We need to transcend our differences, overcome our defenses and prejudices, and develop a deep respect for diversity. Dr. Peck claims we must develop a sense of community or interconnectedness with one another if we are to accomplish our best and ensure our survival as human beings. As illustrated in Figure 15.5, the opposite of this perspective or win/lose independence is often experienced on the road, fueling "road rage" and contributing to numerous vehicle crashes and fatalities.

It seems intuitive that building a sense of community or belongingness among our coworkers will improve OSH. Injury prevention requires interpersonal observation and feedback, and for this to happen, people need to adopt a collectivistic "win/win" perspective instead of the individualistic "win/lose" orientation common in many work settings, including my own University culture. A sense of belonging and interdependency leads to interpersonal trust and AC4P behavior—essential features of a TSC.

In my numerous group discussions with employees on the concept of belongingness, someone inevitably raises the point that a sense of belonging or community at their plant has decreased over recent years. "We used to be more like family around here" is a common theme. For many

FIGURE 15.5 A win/lose independent perspective makes vehicle travel more risky.

> We use more rewards than penalties with *family* members.
>
> We don't pick on the mistakes of *family* members.
>
> We don't rank one *family* member against another.
>
> We brag on the accomplishments of *family* members.
>
> We respect the property and personal space of *family* members.
>
> We pick up after other *family* members.
>
> We correct the at-risk behavior of *family* members.
>
> We accept the corrective feedback of *family* members.
>
> We are our brothers/sisters keepers of *family* members.
>
> We actively care because they're *family*.

FIGURE 15.6 Incorporating an actively caring *family* perspective in an organization will help to cultivate a Total Safety Culture.

companies, growth spurts, continuous turnover—particularly among managers—or "lean and mean" cutbacks have left many employees feeling less connected and trusting.

Figure 15.6 lists a number of special attributes prevalent in most families where interpersonal trust and belongingness are often optimal. We are willing to actively care in special ways for the members of our immediate family. The result is optimal trust, belonging, and AC4P behavior for the safety and health of our family members. Following the guidelines in Figure 15.6 enables the achievement of a TSC. In other words, following the principles listed in Figure 15.6 would develop interpersonal trust and belongingness among members of your "corporate family" and lead to the quantity and quality of AC4P behavior expected among family members—at home and at work.

15.2 ACTIVELY CARING AND EMOTIONAL INTELLIGENCE

The same person-states described here as influencing people's willingness to actively care for the safety and health of others also reflect a most important kind of human dynamic—emotional intelligence (EQ). How important is EQ? Well, it is probably much more responsible for our successes and failures in life than mental capacity, or the "intelligence quotient" (IQ) measured by standard IQ tests. From his comprehensive review of the research, Daniel Goleman (1995) concluded that, "At best, IQ contributes about 20 percent to factors that determine life success, which leaves 80 percent to other factors" (p. 34).

Goleman shows convincing evidence that a majority of the other factors contributing to personal achievement can be associated with EQ or one's ability to:

- Remain in control, resilient, and optimistic following personal failure and frustration.
- Understand and empathize with other people and work with them cooperatively.

In his seminal book, Gardner (1993) refers to the first ability as "intrapersonal intelligence" and the second as "interpersonal intelligence." We show intrapersonal intelligence when we keep

FIGURE 15.7 Interpersonal communication can affect intrapersonal communication, and vice versa.

our negative emotions (including frustration, anger, sadness, fear, disgust, and shame) in check and use our positive emotions or moods (such as joy, passion, love, optimism, and surprise) to motivate constructive action. The driver in Figure 15.7 is attempting to control his negative emotions elicited by an unfriendly interpersonal communication.

Please note the connection between the EQ concept and the AC4P model discussed previously (i.e., Figure 15.1). Each of the AC4P person-states—self-esteem, empowerment, optimism, and belonging—reflect an aspect of EQ, as conceptualized by Goleman (1995, 1998).

Consider also that actively caring for safety increases EQ in ourselves and in others. In other words, when we help people avoid taking a calculated risk in order to achieve a delayed and remote positive consequence (avoiding an injury), we increase this special human capacity in ourselves. When those people willingly follow our OSH advice and give up the efficiency, comfort, or convenience of an at-risk shortcut, they are enhancing this intelligence dynamic within themselves (cf. Geller, 2023).

15.2.1 Safety, Emotions, and Impulse Control

OSH leaders need to develop EQ in themselves and among others. Think about the range of emotions that come into play as we struggle to improve workplace safety and health. We need: 1) curiosity to assess objectively the impact of our OSH interventions, 2) persistence to continue successful programs in the face of active resistance, 3) flexibility to try new approaches, 4) resilience to bounce back after failure, and 5) passion to try again.

Achieving the vision of an injury-free TSC requires awareness and control of our own emotions, as well as the ability to assess, understand, and draw on the influence of other people's emotions. This requires empathic and persuasive communication skills (interpersonal EQ), as well as self-confidence, self-motivation, self-esteem, and optimism (intrapersonal EQ) to develop and implement new interventions for continuous OSH improvement.

Perhaps we can increase personal responsibility for OSH by helping people understand the fundamental emotional issue at the root of all OSH-related interventions. Safety requires impulse control under the most difficult circumstances. We ask employees to do things that are uncomfortable or inconvenient in order to avoid a negative consequence that seems remote and improbable. That takes a special kind of EQ, both from you as a safety leader and from the employees with whom you are working.

15.2.2 Nurturing EQ

The relevance of EQ for improving OSH is obvious, right? Safety leaders need to remain self-motivated and optimistic (intrapersonal EQ) in their attempts to prevent injuries, and much of their success depends upon their ability to facilitate engagement, empowerment, and win/win cooperation among those who could be injured (interpersonal EQ). However, it is easy for safety leaders to get discouraged and frustrated, because so often safety seems to take a back seat to seemingly more immediate demands like meeting production quotas and quality standards.

When we ask people to actively care for safety and health, we are asking them to give up a powerful immediate reward—the ease, speed, or comfort they get from at-risk behavior. In return for extra effort, we promise a bigger reward—they will prevent personal injury or perhaps reduce the possibility that a coworker will be injured. Unfortunately, this delayed reward might not seem credible. People have learned that they can get away with at-risk behavior, and many people have not made the connection between their own behavior and the reduction of injuries among others.

15.3 IN SUMMARY

In this chapter, we continued to develop an understanding of AC4P behavior as it relates to injury prevention. A person-based approach was emphasized by considering dispositions or person-states potential determinants of AC4P behavior, including their emotional intelligence or EQ.

Six person-states were proposed as influencing people's willingness to actively care—self-esteem, self-efficacy, response-efficacy, outcome expectancy, optimism, and a sense of belonging. Each of these person variables has a prosperous research history in psychological science, and some of that research relates directly to the AC4P model. Research that tested relations between particular person-states and actual behavior has generally supported the AC4P model.

A particularly important question is whether the AC4P person-states are both antecedents and consequences of a caring act. It seems intuitive that performing an act of kindness that is effective, accepted, and appreciated could increase the helper's self-esteem, empowerment, optimism, and sense of belonging. This, in turn, should increase the probability of more AC4P behavior. In other words, one act of caring, properly appreciated, should lead to another . . . and another AC4P behavior. A self-supporting AC4P cycle is likely to occur.

The increasingly popular and research-supported concept of EQ relates directly to the AC4P model and to improving safety at work, at home, and on the road. Each of the person-states in the AC4P model reflects EQ, and when people go beyond the call of duty to actively care for the safety or health of others, they build EQ in themselves and in the people they help. Thus, it is crucial to get AC4P behavior initiated and accepted among a large group of individuals. This critical challenge is addressed in the next chapter.

16 Increasing Occurrences of AC4P Behavior

This chapter integrates principles and procedures from previous chapters to address the most critical question regarding the achievement of a Total Safety Culture: How can we increase occurrences of AC4P behavior throughout a work culture? Some conditions and interpersonal techniques facilitate this behavior indirectly by benefiting self-esteem, empowerment, and/or a sense of belonging. Other procedures can boost actively caring directly with certain activators and consequences. In addition, the social influence principles of reciprocity and consistency can be applied to increase occurrences of AC4P behavior. Thus, this chapter shows you how to increase the likelihood that people will go beyond their normal routines to actively care for safety.

In a Total Safety Culture . . . people "actively care" on a continuous basis for safety.

That quotation from my 1994 article in *Professional Safety* (Geller, 1994, p. 18) reflects the ultimate vision of a TSC—a culture where everyone periodically goes beyond his or her personal routine for the safety and health of others. To meet this challenge, we need to find ways to increase the occurrence of AC4P behaviors.

So, how do we do this at work, in our homes, and throughout the community? I hope you can already provide some answers after reading Chapters 14 and 15, where some ways to encourage AC4P behavior were revealed. In this chapter, we elaborate on earlier suggestions and add more. These intervention approaches can be classified as either indirect or direct.

Indirect strategies for facilitating AC4P behavior follow from the theory and principles discussed in Chapter 15. The AC4P model supported by research proposes that certain person-states within people increase their willingness to look beyond self-interests and consider the safety and wellbeing of others. Thus, conditions and procedures that increase these person-states will indirectly increase the amount of actively caring for safety among people. More direct ways to increase occurrences of AC4P behavior can be derived from social influence principles and applied behavioral science. These are discussed in the latter part of this chapter.

16.1 ENHANCING THE AC4P PERSON-STATES

Sometimes at seminars and workshops, I have heard participants express concern that the AC4P person-state model might not be practical. "These concepts are too soft or subjective" is a typical reaction. Employees accept the behavior-based coaching approach because it is straightforward, objective, and clearly applicable to the workplace. However, person-based concepts like self-esteem, empowerment, and belongingness appear ambiguous, "touchy-feely," and difficult to deal with. "The concepts sound good and certainly seem important, but how can we wrap our arms around these 'warm fuzzies' and use them to prevent injuries?"

Those AC4P person-states are more difficult to define, measure, and manage than behaviors. That is why I said early on in this text that it is more cost effective to target behaviors first. Whenever the ultimate outcome is behavior change, it is usually most efficient to deal directly with behaviors. However, it is always important to consider people's feelings when designing behavior-change interventions. That is why I recommend against using negative consequences whenever possible, and why I offer ways to design intervention strategies that account for subjective person-states like commitment, ownership, and engagement.

Increasing Occurrences of AC4P Behavior

After introducing the AC4P person-states at my workshops, I have often divided participants into discussion groups. I asked group members to define events, situations, or contingencies that decrease and increase the particular person-state assigned to their group. Then, I ask the groups to derive simple and feasible action plans to increase their assigned person-state. That assignment promoted personal and practical understanding of the concept. Let's take a look at what my workshop participants have derived regarding each of these person-states.

16.1.1 Self-Esteem

Factors consistently mentioned as shaping self-esteem include communication techniques, positive and negative consequence contingencies, and leadership styles. Participants suggested a number of ways to build self-esteem, including:

- Provide opportunities for personal learning and peer mentoring.
- Increase recognition for desirable behaviors and individual accomplishments.
- Solicit and follow up on a person's suggestions.

Communication strategies. Figure 16.1 lists 11 words beginning with the letter "A" that imply a specific verbal technique for increasing a person's self-esteem. Each "A" word suggests a slightly different communication approach, from stating simple words of agreement, admiration, appreciation, and approval to acknowledging the achievement and individual creativity of others through active listening and interpersonal praise. It is also a good idea to argue less and avoid criticizing. Arguments waste time and usually promote a win/lose perspective, and criticism always does more harm than good. Indeed, "constructive criticism" is an oxymoron. Criticism can never be constructive.

When offering corrective feedback, it is critical to be a patient, active, and empathic listener. Allow the person to make excuses, and do not argue about them. Resist the temptation. Giving excuses is just a way to protect one's self-esteem, and it is generally a healthy response.

- *Accept* - appreciate diversity
- *Actively Listen* - with verbal and nonverbal behavior
- *Agree* - with verbal and nonverbal behavior
- *Admire* - "attractive dress," "nice tie"
- *Appreciate* - "please" (activator) and "thank you" (consequence)
- *Acknowledge* - the achievements of other
- *Approve* - praise for good behavior
- *Ask* - for feedback, advice, opinions, etc.
- *Attend* - lend a helping hand
- *Avoid Criticizing* - it won't be accepted anyway
- *Argue Less* - arguments are win/lose situations

FIGURE 16.1 Apply "A" strategies to increase others' self-esteem.

Remember, you already made your point by noting the error and suggesting ways to avoid that same mistake in the future (as discussed in Chapter 12 on coaching). Leave it at that.

16.1.2 SELF-EFFICACY

As explained in Chapter 15, self-efficacy is more situation specific than self-esteem, and so it fluctuates more readily. Job-specific feedback should actually affect only one's perception of what is needed to complete a particular task successfully. It should not influence feelings of general self-worth. Keep in mind, though, that repeated corrective feedback can have a cumulative effect, chipping away at an individual's perceived self-worth.

Achievable tasks. What makes for a "can do" attitude? Personal perception is key. A supervisor, parent, or teacher might believe s/he has provided everything needed to complete a task successfully. However, the employee, child, or student might not think so. Hence, the importance of asking, "Do you have what you need?" You are checking for feelings of self-efficacy. This is easier said than done because people often hesitate to admit they are incompetent. Really, who likes to say, "I can't do it?" Instead, we try to maintain the appearance of self-efficacy.

Figure 16.2 reflects the need to focus on "small wins" when assigning tasks and communicating performance feedback. Of course, the kind of situation depicted in Figure 16.2 requires one-on-one observation and feedback during which the individual's initial competencies are assessed. Then, successively more difficult performance steps can be designed for the learner. The key is to reduce the probability that the learner will make an error and feel lowered effectiveness or self-efficacy. Celebrating small-win accomplishments builds self-efficacy and enables support from the self-fulfilling prophecy.

Focus on the positive. Many of the strategies presented here for improving behaviors and person-states include a basic principle: Focus on the positive. Whether attempting to build your own self-efficacy or that of others, success needs to be emphasized over failure. Thus,

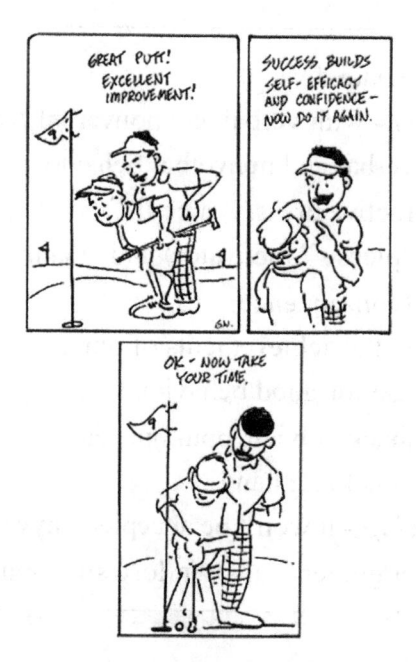

FIGURE 16.2 Small, successive steps to success build self-efficacy.

whenever you have the opportunity to teach others or to give them feedback, you should look for small-win accomplishments and give genuine approval before commenting on ways to improve. Unfortunately, this approach is easier said than done.

Failures are easier to spot than successes. Errors stick out and disrupt the flow. That is why most teachers give rather consistent negative attention to students who disrupt the classroom, while giving only limited positive attention to students who remain on task and go with the flow. However, considering the impact that consequences have on internal person-states (especially self-efficacy and self-esteem), a positive consequence (like praise and social approval) is always preferable to a negative consequence (like criticism or ridicule).

16.1.3 Personal Control

Employees at my seminars on actively caring for safety have listed a number of ways to increase perceptions of personal control, including:

1. Set short-term goals and track progress toward long-term accomplishment.
2. Offer frequent supportive and corrective feedback for process activities rather than only for outcomes.
3. Provide opportunities to set personal goals, teach others, and chart "small wins."
4. Teach employees basic behavior-change intervention strategies (especially feedback and recognition procedures).
5. Provide people time and resources to develop, implement, and evaluate intervention programs.
6. Show employees how to graph daily records of baseline, intervention, and follow-up data.
7. Post response feedback graphs of group performance.

Figure 16.3 illustrates humorously a personal control perspective. Obviously, this is an extreme and unrealistic scenario, but wouldn't it be nice if people would attempt to take personal control

FIGURE 16.3 Sometimes we try extra hard to exert personal control.

of safety issues at their industrial sites with the same passion and commitment some individuals have for their golf game? I believe differences in perceived personal control for safety vs. golf are largely due to contrasting scoring procedures.

Suppose you could not receive direct and immediate feedback about your golf game. That is, each time you hit a golf ball you wore a blindfold and could not see where the ball landed. Even when putting on the greens, you are blindfolded and cannot tell whether your ball goes into the cup. Imagine also that you do not receive a score per hole or per game. However, you do receive negative feedback whenever your ball lands in a sand trap. Under such circumstances, would you feel "in control" of your golf game? Would you attribute balls hit into sand traps to personal control or just bad luck? Would you continue playing golf or give it up for an activity in which you can experience greater personal control?

Of course, the golf scenario I asked you to imagine is far-fetched, but is that the way it is for OSH at many industrial sites? The primary evaluation tool used to rank companies and determine performance appraisals and bonuses is an outcome number (such as total recordable injury rate) that is quite remote from the daily plant processes over which people have control. Without a scoring system that focuses on controllable processes (as discussed earlier), safety will be viewed as beyond personal control. In such workplaces, an injury is just "bad luck," analogous to hitting a golf ball in a sand trap while blindfolded. With that mindset the unfortunate term "accident" is appropriate—"We had no control over the contributing factors and the injury could not have been prevented."

Obviously, we cannot have complete control over all factors contributing to an injury. That is why it is wrong to say "*all*" injuries are preventable." However, there is much we can do within our own domain of influence, and we can prepare for factors outside our personal control. Thus, we take an umbrella to the golf course in case it rains, and we wear PPE in case we are exposed to risks beyond our domain of personal control. Likewise, we protect our children from events beyond their control, as reflected in Figure 16.4.

FIGURE 16.4 We can't control everything!

16.1.4 Optimism

As discussed in Chapter 15, optimism results from thinking positive, avoiding negative thoughts, and expecting the best to happen. Anything that increases our self-efficacy should increase optimism. Also, if our personal control is strengthened, we perceive more influence over our consequences. This gives us more reason to expect the best. Again, we see how the person-states of self-efficacy, personal control, and optimism are clearly intertwined. A change in one will likely influence the other two.

Recall from Chapter 15 that simple events like finding a dime in a coin return, receiving a cookie, listening to soothing music, and being on a winning football team are sufficient to boost optimism and a willingness to actively care. It is not necessary for a person to perceive personal control over a pleasant consequence for that consequence to build optimism and facilitate the occurrence of AC4P behavior.

16.1.5 Belonging

Here are some common proposals given by my seminar discussion groups for creating and sustaining an atmosphere of belonging among employees:

- Decrease the frequency of top-down directives and "quick-fix" programs.
- Increase team-building discussions, group goal-setting and feedback, as well as group celebrations for both process and outcome achievements.
- Use self-managed or self-directed work teams.

When groups are given control over important matters like developing an observation-and-feedback coaching process or a behavior-based incentive/reward program, feelings of both empowerment and belongingness can be enhanced. When resources, opportunities, and talents enable team members to assert, "We can make a difference," feelings of belonging occur naturally. This leads to synergy, with the group achieving more than could be possible from participants working independently, as illustrated in Figure 16.5.

FIGURE 16.5 Higher goals can be reached through synergy.

FIGURE 16.6 Win/lose competition inhibits teamwork and synergy.

Figure 16.6 depicts an obstacle to win/win synergy that I have witnessed all too often, and seems to be stronger than ever in politics these days. The "we/they" attitude concomitant with traditional management/labor differences often makes for dysfunctional work groups. It seems some unions attempt to justify their existence by focusing on disagreement, conflict, and mistrust between management and labor. For its part, management often supports this "we/they" split with an alienating communication style that asserts its ultimate power and control. I have seen management memos, for example, that might have been well-intentioned but were written in top-down, control language that reflected an adult talking to a child.

By actively caring for safety, and by enhancing self-esteem, empowerment, and belonging, we can bring down the "us vs. them" walls that entrap a work culture. AC4P behavior spreads mutual trust and interdependence throughout the culture. In a TSC, everyone benefits from each individual's efforts.

Does AC4P imply the elimination of unions? No, but it might suggest altered visions and mission statements for organized labor groups. Labor unions can certainly help enhance the person-states that facilitate AC4P behavior. To do this, they need to work with management from a win/win perspective that appreciates interdependence and the power of synergy.

At the same time, managers need to relinquish their hold on the "control buttons" of operations and processes that workers could manage themselves, perhaps through self-directed work teams. A truly "empowered workforce" is one that is trusted by managers and supervisors to get the job done without direct supervision. Obviously, this cannot happen overnight, but a solid foundation is laid when the AC4P person-states are strengthened.

16.2 DIRECTLY INCREASING OCCURRENCES OF AC4P BEHAVIOR

You can treat AC4P behavior just like any other target behavior. Many interventions that increase the occurrence of safe work behaviors can be used to boost the frequency of AC4P behavior. The four chapters in Part 4 covered principles and procedures for directly influencing behavior. Recall that the techniques were classified as activators and consequences, with activators considered directive or instructional, and consequences being motivational. Let's review that discussion as it applies to AC4P behavior.

16.2.1 Education and Training

There is some evidence that educating people about the barriers inhibiting AC4P behavior will increase acts of caring. The impact of education should be even more dramatic if some of the concepts discussed in Chapter 14 are taught. It seems particularly useful to explain that AC4P behavior can be proactive or reactive, direct or indirect, and focused on the environment, internal person-states, or external behavior. Also, it should be taught that direct, proactive, and behavior-focused AC4P is most challenging, yet most useful for the prevention of unintentional injuries.

16.2.2 Consequences for Actively Caring

Relevant rewards should increase AC4P behavior. This comes from the basic operant-learning principle that the recurrence of behavior increases following positive consequences. However, it is perhaps even more important not to respond negatively when observing a caring act. A negative reaction could make that person avoid a subsequent opportunity to actively care for safety. A similar point was made when discussing safety coaching in Chapter 12. A person's reaction to a safety coach could determine whether or not that person will go out of his or her way to coach again for safety.

The "Actively Caring Thank-You Cards" described in Chapter 11 epitomize a rather generic technique to boost occurrences of AC4P behavior in the workplace. Employees used those cards to thank their colleagues for going beyond the call of duty for the safety of others. Some cards included a space for the intervention agent to define the act of caring. Some cards could be exchanged for inexpensive rewards, like a beverage in the company cafeteria. Other cards had peel-off stickers for public display. In some applications, each card was worth some amount of money (e.g., $1.00) toward corporate contributions to local charities.

16.2.3 The Reciprocity Principle

Some sociologists, anthropologists, and moral philosophers consider reciprocity a universal norm that motivates a good deal of positive interpersonal behavior. Simply put, people are expected to help those who have helped them. You can expect people to comply with your request if you had previously done a favor for them. This is the principle behind Stephen Covey's (1989) claim that we need to make deposits in another person's emotional bank account before making a withdrawal.

Have you ever felt a little uncomfortable after someone did you a favor? I certainly have. I interpret my discomfort as the reciprocity principle in action. Another person's kind act makes me feel pressured to respond in kind. What does this mean for OSH? I think it means we should look for opportunities to go out of our way for other people's safety. By doing this, we increase the likelihood that those individuals will help us when we need them.

Gifts aren't free. Has someone snared your attention to hear a sales pitch after giving you a free gift? Have you ever felt obligated to contribute to a charity after receiving gummed individualized address labels and a stamped envelope for your check? Ever purchase food in a supermarket after eating a free sample? Do you feel obliged to buy something after using it for a ten-day "free" trial period?

If you answer "yes" to any of those questions, it is likely you have been influenced by the reciprocity principle. Many marketing or sales-promotion efforts count on this "free sample" gimmick to influence purchasing behavior or a financial contribution. Does this justify distributing free safety gifts, such as pens, tee-shirts, caps, cups, and other trinkets? Yes, to some extent, but you have to consider perceptions. How special is the gift? Was the gift given to a select group of people, or was it distributed to everyone? Does the gift or its delivery represent significant sacrifice in money, time, or effort? Can it be purchased elsewhere, or does its safety slogan make it special?

A "special" safety gift—as perceived by the recipient—will inspire more acts of caring in response. Remember, too, that the way a gift is bestowed can make a big difference. The labels and slogans linked with a "gift" can influence the amount and kind of responsive action. If the gift is presented to represent the actively caring for safety leadership expected from a "special" group of workers, a certain type of reciprocity is activated. People will tell themselves they are considered AC4P leaders, and they need to justify this label by going beyond the call of duty for the safety and health of others. If they have learned about the various categories of AC4P behavior, they know the best way to lead is by taking action that is direct, proactive, and focused on supporting safe behavior or correcting at-risk behavior.

16.2.4 Commitment and Consistency

Robert Cialdini refers to commitment and consistency as an influence mechanism lying deep within us that direct many of our actions. It reflects our motivation to be, and appear to be, consistent. "Once we make a choice or take a stand, we will encounter personal and interpersonal pressures to behave consistently with that commitment" (Cialdini, 2001, p. 53). Cialdini suggests the pressure comes from three basic sources.

- Society values consistency within people.
- Consistent conduct benefits daily existence.
- A consistent orientation allows us to take shortcuts when processing information and making decisions. We do not have to stop to consider everything involved; instead we fall back on our prior commitment or decision and act accordingly to remain true to ourselves.

Public and voluntary commitment. The "Safe Behavior Promise Card" described in Chapter 10 derives its influential power from the commitment and consistency principle. When people sign their name to a promise card they commit to behaving in a certain way. Then, they act in ways consistent with their commitment.

Commitments are most effective, or influential, when they are visible; require some amount of effort; and are perceived to be voluntary, not coerced. It makes sense, then, to have employees state a public rather than a private commitment to actively care for safety and to have them sign their name to a promise card rather than merely raise their hand. It is critically important for those making a pledge to believe they did it voluntarily. In reality, decisions to make a public commitment are dramatically influenced by external activators and consequences, including peer pressure. However, if people sell themselves on the idea that they made a personal choice, consistency is likely to follow the commitment.

Figures 16.7 and 16.8 illustrate a public commitment intervention implemented at a safety seminar for supervisors, safety leaders, and maintenance personnel of Delta Airlines. After giving a keynote address on the concept of actively caring for the safety and health of others, I signed my name to a "Declaration of Interdependence" as a symbol of commitment to actively care for the safety of others (Figure 16.7). Then, I urged the audience to follow suit.

The social context was partly responsible for the substantial number of individuals signing the declaration (Figure 16.8), which was later prominently displayed at the employees' worksite. The public and voluntary nature of the commitment request contributed to the effectiveness of this exercise to activate awareness and the subsequent development of OSH-related goals and action plans.

Foot-in-the-door: Start small and build. This strategy follows directly from the commitment and consistency principle. To be consistent, a person who follows a small request will likely comply with a larger request later. During the Korean War, the Chinese communists used this technique on American prisoners by gradually escalating their demands, which started with a few harmless requests. First, prisoners were persuaded to speak or write trivial statements.

FIGURE 16.7 The author signs a "Declaration of Interdependence."

FIGURE 16.8 Delta Airlines employees sign a "Declaration of Interdependence."

Subsequently, they were urged to copy or create statements that criticized American capitalism. Eventually, the prisoners participated in group discussions of the advantages of communism, wrote self-criticisms, and gave public confessions of their wrong-doing.

Research has found this "start small and build" strategy succeeds in boosting product sales, monetary contributions to charities, and blood donations. However, this "foot-in-the-door" technique only works when people comply with the first small request. If a person says "No" right away, s/he might find it even easier to resist subsequent, more important requests. So, if your first call for AC4P behavior is shot down, you did not start small enough. Be prepared to retreat to something less demanding and build reciprocity from there.

16.3 IN SUMMARY

The AC4P information reviewed in this chapter integrated principles and procedures from prior chapters. Continuous safety improvement leading to a TSC requires people to actively care for the safety of others as well as themselves. Evidence-based procedures to increase the frequency of AC4P behavior throughout a culture were elucidated. Some of those techniques indirectly facilitate AC4P behavior by benefiting the person-states that facilitate one's willingness to actively care. Other strategies target AC4P behaviors directly.

Indirect strategies are deduced from the AC4P model introduced in Chapter 15. Any procedure that increases a person's self-esteem, perception of empowerment, optimism, and/or a sense of belonging or group cohesion will indirectly benefit AC4P behavior. A number of communication techniques enhance more than one of those person-states simultaneously, particularly listening with empathy to others for their feelings and giving them genuine praise for their accomplishments. There are barriers to focusing on the positive, and those were considered with the hope that awareness will help overcome the obstacles.

The behavior-improvement techniques detailed in Part 4 of this text—from setting SMART goals and signing promise cards to offering soon, certain, and positive consequences for desirable behaviors—can be used to enhance those AC4P person-states. They can also directly increase AC4P behavior. Education and direct participation in AC4P projects foster the occurrence of AC4P behaviors.

The interpersonal influence principles of reciprocity and consistency were introduced as they apply to our everyday decision-making and behaviors. The consistency principle is behind the success of a safe-behavior promise card and the foot-in-the-door technique ("start small and build"). These principles and evidence-based practical strategies can be applied to directly increase the frequency of AC4P behavior.

Part Six

Putting It All Together

17 Promoting High-Performance Teamwork

Everyone talks about teamwork, but not everyone gets the best from their teams. Teamwork does not come naturally. This chapter explains why and what we can do about it. Principles and practical procedures are offered for initiating and sustaining productive teamwork. The functions of seven different safety teams are described. Each of those teams contributes to improving the human dynamics of OSH, as taught in this text. Each team depends on the output of other teams in order to optimize the system and cultivate a Total Safety Culture.

> My responsibility is to get my twenty-five guys playing for the name on the front of their uniform and not the name on the back.
>
> **(Tommy Lasorda)**

Imagine a workplace where everyone coaches each other about the safest way to perform a job—a workplace in which actively caring for other people's safety is a natural part of the everyday routine. When people depend on each other in this way to improve OSH, they understand teamwork. They have an interdependent mindset and realize the true meaning of synergy. For such individuals, TEAM means: **T**ogether **E**veryone **A**chieves **M**ore.

Reaching this level of teamwork does not come easy. Consider how we have been raised. "Be independent," we are told. We continually compete with other individuals to get ahead, whether in school, at work, or at play. Remember these common slogans: "Nice guys finish last," "The squeaky wheel gets the grease," and "You have to blow your own horn." A win/lose, me-first mindset is integrated in so many situations, from the grades we get in school to salary promotions at work, and it is promoted nonstop on social media.

This chapter is about win/win teamwork. It builds on the principles of behavior-based safety presented in Part 3, the intervention tools from applied behavioral science detailed in Part 4, and the concepts of group belongingness and interdependence discussed in Part 5.

17.1 CULTIVATING HIGH-PERFORMANCE TEAMWORK

Let's consider the main phases of teamwork, from start to finish, and see how each relates to the development of optimum group performance for continuous OSH improvement. In this chapter, practical answers are provided to questions such as, "How can we establish a successful safety team?" and "Once we have a safety team, what can we do to make it more effective?"

Figure 17.1 outlines seven consecutive phases of teamwork, from selecting team members to disbanding or renewing the team. Those are the basic steps of teamwork as discussed by leading team-building trainers and consultants. Let's examine each of these steps in more detail as they relate specifically to OSH.

17.1.1 Selecting Team Members

The first crucial step in successful OSH teamwork is to select the right people for your team. Someone is ultimately responsible for choosing team members. For OSH, this is often the safety director or the person responsible for maintaining injury reports and lost-time records. However,

> 1. **Select the Right Team Members**
>   - Understand and appreciate behavior-based safety.
>   - Commitment, interpersonal interest, communication, and caring.
> 2. **Clarify the Assignment**
>   - State general mission or purpose.
>   - Specify resources, authority, and accountability.
>   - Get acquainted.
>   - Develop understanding of TEAM, interdependency, and synergy.
> 3. **Establish a Team Charter**
>   - Write a mission statement.
>   - Set ground rules.
>   - Define deliverables and accountability.
>   - Specify budget details and direct reports.
>   - Assign standard team roles.
> 4. **Develop Action Plan**
>   - Set goals with SMARTS.
>   - Assign task responsibilities.
>   - Develop time lines.
> 5. **Engage in the Process**
>   - Conduct productive team meetings.
>   - Use brainstorming and consensus-building.
>   - Give each other supportive and corrective behavior-based feedback.
> 6. **Evaluate Team Performance**
>   - Recognize process results.
>   - Document product results.
>   - Celebrate accomplishments.
> 7. **Disband, Restructure, or Renew the Team**

FIGURE 17.1 Follow seven steps for team success.

in some cases, it is advantageous for a small committee of safety champions representing a cross-section of the workforce to select *potential* members of a safety team. I say "potential" because it is important for membership to be voluntary. So, a safety champion or selection committee should come up with a list of people to approach one on one and ask if they would be willing to serve on a particular safety team.

So, what kinds of people should you look for as potential members of an OSH team? Perhaps, first and foremost, the candidate should be committed to safety. Has the individual done something recently to indicate personal concern for the safety or health of a coworker? Perhaps s/he turned in a comprehensive close-call report in a work culture where such reports are rare, or maybe the employee was injured recently and has given testimony to a renewed regard for safety initiatives.

Besides demonstrating special commitment to OSH, the best team members also have other qualities. They have interpersonal skills, they like to work with other people, they communicate well, they actively listen and speak with passion, and they are willing to actively care for the safety and health of others.

17.1.2 Clarify the Assignment

From the start, it is important for the team members to understand their assignment. They need to know the overall mission of the team and the resources available to accomplish it. They also need to understand their authority with regard to the mission. For example, they need to realize the degree of control the team has over the consequences of their decisions. Will any other authority

Promoting High-Performance Teamwork

influence the outcome of the team's decisions? In other words, to what extent is the team truly empowered to carry out the processes needed to accomplish its mission?

The true purpose of a team needs to be discussed. Why is the assignment given to a team instead of individuals working alone? What are the advantages of using a team? Own up to the fact that the start-up process will take some time, and some might think one dedicated person would be more efficient. In the long run, however, teamwork will be more effective. Explain the TEAM acronym—Together Everyone Achieves More.

Besides explaining the general mission and the TEAM concept, the team facilitator should also provide opportunities for the participants to get acquainted. Introductions could be initiated with a statement like, "Let's get to know each other better by each person stating your name, your job, and your general expectations, if any, about our team assignment." Later, you might also ask each team member to say something about safety. "Tell us about your personal interest or commitment to occupational safety and health."

17.1.3 Establish a Team Charter

During the prior "getting acquainted" phase, the mission was given as a general description of the team's assignment. Now it is time to write a formal mission statement that articulates the overall purpose of the team, defines the ground rules for team meetings, addresses budget issues, specifies what the team will produce (its deliverables), and assigns various team roles. Some standard team roles are as follows, although sometimes one person assumes more than one role.

- Team leader: provides direction and obtains outside resources.
- Team facilitator: keeps meetings focused and prompts total participation.
- Team administrator: handles various administrative duties like distributing reports, networking with outside individuals and groups, and reminding members of team meetings.
- Treasurer: tracks the input and output of finances.
- Reporter: documents and distributes the meeting agenda and minutes.

It is vital that each team member understand and affirm the mission statement, deliverables, ground rules, accountability system, and team role assignments. Therefore, a "team charter" is developed through consensus-building, which is the opposite of top-down decision-making. It is not the same as negotiating, calling for a vote and letting the majority win, or working out a compromise between two different sets of opinions. The team facilitator in Figure 17.2 is not likely to cultivate consensus.

Building consensus. Negotiating, voting, or compromising comes across as win/lose and decreases the interpersonal trust needed for high performance teamwork. A majority of the team members might be pleased, but other members will be discontented and might actively or passively resist involvement. Even the "winners" could feel lowered interpersonal trust. "We won this decision, but what about next time?" Without everyone's buy-in for a group decision, the teamwork will be less synergistic than possible.

So how can group consensus be developed? How can the outcome of a heated debate on ways to solve a problem be perceived as a win/win solution everyone supports instead of a win/lose compromise or negotiation? Practical answers to these questions are easier said than done. Consensus-building takes time, energy, and patience. It requires open and frank conversation among all team members.

The scenario depicted in Figure 17.3 is unacceptable. All participants must be willing to state their honest opinion without fear of ridicule or reprisal. It takes a good team facilitator to make this happen. S/he needs to solicit the opinions of everyone throughout the team meeting.

There is no quick fix to do this. It requires plenty of interpersonal communication, including straightforward opinion sharing, intense discussion, emotional debate, proactive listening,

FIGURE 17.2 Body language speaks louder than words.

FIGURE 17.3 Consensus-building requires frank and open communication.

careful evaluation, methodical organization, and systematic prioritizing. On important matters, however, the outcome is well worth the investment. When you develop a solution or process every potential participant can get behind and champion, you have cultivated the degree of interpersonal trust and ownership needed for total involvement. Involvement, in turn, builds

> 1. **Everyone participates.**
>   Active participation comes with the territory. This means always being prepared for team meetings and problem-solving discussions.
> 2. **No barbs or put-downs.**
>   Team members show respect for each other. They don't say anything that could hurt someone's feelings or limit the involvement of others.
> 3. **Every idea counts.**
>   Team members listen with respect to everyone's opinion regardless of how silly it might seem at first. The strangest sounding idea can be the seed for creative invention.
> 4. **Strive to be completely informed.**
>   Team members actively listen during team meetings to know exactly what's going on. They openly ask questions about anything they don't fully understand.
> 5. **Follow through on commitments and meet deadlines.**
>   Keeping promises builds interpersonal trust and assures team progress. This includes showing up for *all* meetings on time.
> 6. **Support team decisions.**
>   Team members voice their concerns during decision-making discussions because in the end they realize they must support a team decision.
> 7. **Think win/win interdependency.**
>   Team members want everyone to win. Synergy depends on everyone contributing individual talents for the good of all.

FIGURE 17.4 Seven ground rules can promote effective team meetings.

personal commitment, more ownership, and then more involvement. The concept of "psychological safety" is very relevant here, as explained in Chapter 19.

Developing ground rules. Total participation of every team member during important discussions is vital for consensus-building. Thus, open and frank discussion should be a ground rule for team meetings, and accepted by everyone. Figure 17.4 lists this and other potential ground rules to consider adopting for increasing the effectiveness of your team meetings. These are general guidelines or rules of conduct on which the team members need to agree and hold each other accountable to follow.

You should not just post the contents of Figure 17.4 and call them "our team ground rules." Rather, you need to discuss the issue of ground rules with team members and get everyone's opinion. Then, use a consensus-building approach to get everyone's acceptance of the final list. You can use the suggestions in Figure 17.4 to "prime the pump" or stimulate discussion. Alternatively, you could just keep the seven items in mind as you facilitate group discussion and look for opportunities to direct comments toward those topics.

Writing a mission statement. You can use this basic consensus-building approach to arrive at a mission statement. In other words, you should have a basic idea of the team's overall mission or purpose. Then, use consensus building to write a mission statement everyone can support. Be sure to explain that the team's mission statement should be a clearly stated purpose that serves to direct and motivate team members. It answers three basic questions:

- What does the team do?
- How does the team do its work?
- Who are the team's customers?

17.1.4 Develop an Action Plan

Keeping in mind the team mission and ground rules, the team now plans how it will proceed. This planning process consists of three primary steps.

- Define specific goals needed to accomplish the mission.
- Decide on a timeline for completing each goal.
- Assign goal-relevant tasks to each team member.

Let's consider some general approaches to taking these steps. The specifics will vary depending upon the OSH mission; the team charter; and the talents, skills, and opinions of the team members. Remember to follow the basic consensus-building process when arriving at goals and assigning tasks.

Define tasks. Process goals imply specific tasks, while outcome goals do not. For example, a process goal: "Complete 100 behavior-based observation-feedback sessions by the end of the month" is quite task specific and stipulates what needs to be done. In contrast, the outcome goal: "Reduce the total recordable injury rate by 50 percent this year" does not suggest any actions or behaviors. Assuming this outcome goal is judged achievable by the team, it is necessary to decide on specific tasks needed to achieve the goal. One of those tasks might be, in fact, to conduct periodic behavior-based observation-feedback sessions throughout a work area. In this case, the process goal: "Achieve 100 observation-feedback sessions" would be needed to attain the outcome of reduced injuries.

It would likely take several process goals to reach a certain injury-reduction outcome goal. Process goals could be derived for attendance of safety meeting, for reporting of close calls and property-damage incidents, for removing environmental hazards, and for engaging in one-on-one OSH-related conversations with coworkers. All of these are tasks that could contribute to reaching an injury-reduction milestone. So, the achievement of an injury-reduction outcome goal is contingent on reaching certain process goals, which are consistent with the mission or purpose of achieving an injury-free workplace.

Assign task responsibilities. The key to successful teamwork is to develop a list of specific tasks needed to achieve team goals; and then, to assign the right persons to take on the various task responsibilities. Also critical, of course, is the setting of appropriate deadlines for each task to be completed. Adding a deadline or completion date to specific assignments results in a SMARTS goal. These goals are SMART, as discussed earlier in Chapter 10, with "S" for specific, "M" for motivational, "A" for achievable, "R" for relevant, and "T" for trackable. The added "S" for SMARTS team goals stands for shared. SMARTS goals are then organized into a timeline for scheduling teamwork.

17.1.5 Make it Happen

After setting SMARTS goals, assigning task responsibilities, and developing a timeline, the real work begins. Process goals have been set and team members are motivated to fulfill their interdependent roles for the sake of their team mission. This is the performing stage of teamwork. Regular team meetings are still needed to keep the process going and to promote continuous improvement. More specifically, team meetings provide opportunities for team members to connect with one another and:

- Hold each other accountable for achieving specific tasks.
- Review project progress and acknowledge achievements of individuals, subgroups, and the team as a whole.
- Discuss problems and entertain corrective action plans.

- Check the timeline and make refinements and additions.
- Plan for next steps and assign new task responsibilities.

Prepare an agenda. A well-planned and well-run meeting starts with an agenda. This keeps the meeting on track and focused, and prevents off-track or time-wasting behavior. A good agenda is neither wordy nor complex. It is simply an outline of items to be covered.

At the start of the meeting, a copy of the agenda should be distributed to all participants. When participants keep an agenda in front of them throughout the meeting, they are apt to keep their comments on track. This also prompts team members to offer their perspectives and recommendations at the appropriate times.

The basic structure of the agenda will not vary much from one meeting to the next. The following components are included in most meeting agendas, and they should occur in the order given here.

- Review the purpose of the team meeting.
- Make any organizational announcements relevant to the team's mission.
- Call for progress reports from team members, including project objectives, accomplishments since the last meeting, and special assistance or resources needed for the next steps.
- Discuss special issues, difficulties, and solutions with a focus on the positive or on examining ways to overcome problems raised.
- Identify what needs to happen next per project or task assignment in order to progress and continuously improve.
- Set the time and date for the next meeting, and offer a preview of critical topics or project reports to be covered.

Manage time well. How often have you heard someone say, "I could get so much more done in a day if I didn't have so many meetings to go to"? Meetings do have the reputation of robbing people of valuable time. So, if time is managed well at your OSH team meetings, everyone will be grateful. More will get accomplished, and people will feel good about the time they gave up. Consider the following suggestions for managing your meeting time effectively.

- Designate a start and stop time, and make sure everyone knows what those are.
- Start the meeting on time, even if everyone has not shown up. This sets the stage for on-time arrivals.
- Stop the meeting on time, even if every agenda item was not covered completely. This sets the stage for efficient use of meeting time.
- Allocate specific time periods for each agenda item and remind participants of these throughout the meeting.
- If breaks are given, state a precise time for participants to return. Start the meeting again at that time, even if everyone has not returned.
- If discussions get long, remind participants of the time remaining and the number of agenda items left.
- Hold meetings prior to lunch time or at the end of the workday in order to provide an incentive to get things done on time.
- Disallow cellular phones or pagers in the meetings or you will set the stage for distraction.

Record minutes. Decisions and assignments made during a team meeting are usually critical for team success, yet they can be easily forgotten. Therefore, it is important for someone to document key events of the team meeting. When team members are confident the designated "recorder" will take good notes, they will not be distracted by their own note-taking behavior.

They can listen attentively and participate actively throughout the meeting. Later, the meeting notes are made available as a permanent record of team progress. This reminds teammates of their accomplishments and their obligations for continual success.

Communicate between meetings. Team members need to support each other when completing their assignments. Do not wait until the next team meeting to inquire about a teammate's progress on a project. Asking others how their specific assignments are going sends the message that you care about their contributions to your team. Taking the time to listen to a report of progress or actually reviewing results from a particular assignment shows your concern and can be a powerful motivator.

It is crucial to give supportive and corrective feedback for specific behaviors related to team assignments. As discussed in Chapter 16, behavior-based feedback is extremely important in directing and motivating desirable behavior. Team members need to be alert to the kinds of behaviors needed from each other in order to have a successful team. They also need to follow certain guidelines for giving effective feedback to their teammates, as detailed earlier in Chapter 12.

17.1.6 EVALUATE TEAM PERFORMANCE

The value of a performance evaluation cannot be denied, when it is done well. In fact, the discussion in the previous section about feedback says it all. Willing workers cannot improve without receiving feedback directly related to their performance, and such feedback can be available through an objective and periodic evaluation process. Evaluation is key to accountability and responsibility, as detailed in the next chapter. Here, a few guidelines are offered regarding the evaluation of OSH teams for optimization.

Why might some readers find Figure 17.5 humorous? What is the problem with many performance appraisals? If performance appraisals were objective, fair, and based on behavior, you

FIGURE 17.5 Subjective, nonbehavioral, and infrequent evaluations are not taken seriously.

would not see any humor in that illustration, right? So, a useful evaluation of team performance needs to be objective, fair, and related to relevant behaviors and conditions.

Your team needs to evaluate its performance periodically in order to assess successive improvements recommended in prior evaluations. Then, it has reason to celebrate its accomplishments. Quality team celebrations (as discussed in Chapter 13) are key to enhancing team cohesiveness and mutual responsibility toward the accomplishment of more SMARTS goals.

It is important to realize that although performance evaluation is listed sixth on the list of successive teamwork steps in Figure 17.1, this topic is inherent in every step. Whether selecting team members, establishing a team charter, or setting goals and assigning task responsibilities, evaluation plays an integral role.

Team members continually evaluate each other's opinions and reactions throughout group discussions in order to arrive at decisions everyone will support. The presentation of project reports at team meeting is essentially an evaluation process. Team members appraise whether a project is progressing as planned and decide whether the timeline needs adjusting when the results call for refinements or additions to the list of task assignments. All of this involves the ongoing study and interpretation of information in order to make the best decisions. This is evaluation in the truest sense of the word.

17.1.7 Disband, Restructure, or Renew the Team

Many books and manuals on teamwork discuss this final stage as the time when members of a work team realize their work is done and adjourn or disband. However, this is rarely the case for OSH teams. The work of those teams is never done. Consider, for example, the seven teams defined in Figure 17.6 that are needed to address comprehensively the human dynamics of OSH.

Specific projects or assignments may come and go, but OSH teams need to work persistently on their general missions in order to achieve continuous OSH improvement throughout

---

Safety Steering Team — oversees the effort of all other teams listed here.
Observation and Feedback Team — develops, implements, evaluates, and refines behavior-based observation and feedback procedures.
Ergonomics Team — conducts periodic audits of workplace settings, evaluates employee suggestions regarding ergonomic issues, and recommends corrective action for environment, behavior, or both.
Incident Analysis Team — conducts fact-finding evaluations of close-call reports and injuries, including behavioral, environment, and person-based factors; and recommends corrective action.
Celebration Team — plans and manages celebration events to recognize process activities and reward achievements of milestones.
Incentives/Rewards Team — oversees the design, implementation, evaluation, and refinement of behavior-based incentive/reward programs to motivate participation in designated safety-improvement activities.
Preventive Action Team — evaluates reports of rule/policy violations, decides whether the violator should be punished, and chooses the penalty.

FIGURE 17.6 Various types of employee teams are needed to improve occupational safety.

a work culture. The membership of these teams will change periodically and team goals will vary, but the challenges of behavior observation and feedback, incident analysis and corrective action, ergonomics analysis and intervention, and behavior-based recognition and celebration will remain. Of course, the methods used to meet those team functions will change and, in fact, they will successfully improve if appropriate evaluation processes are implemented.

Safety teams will learn to work more effectively and efficiently over time, but they will need to keep working. Even when your workplace becomes injury free, the safety teams listed in Figure 17.6 are needed to maintain this enviable situation. Thus, this final step for safety team success should be considered as restructuring or renewing, not disbanding.

Restructuring. Restructuring could mean a change in focus, in team membership, or in the methods the team uses to accomplish its mission. For example, after observation-and-feedback teams are in operation throughout your work culture, the Safety Steering Team changes its focus from promoting and training to advising and maintaining. In other words, after employee teams get involved in behavior-based coaching, the challenge becomes one of sustaining the process. The prime issue changes from "How can we teach coworkers behavior-based coaching procedures and convince work teams to use critical behavior checklists on a regular basis?" to "How can we keep work teams motivated to keep their behavior-based observation-and-feedback process going?"

In the beginning, the efforts of the Incentive/Reward Team might focus on convincing management and coworkers to substitute behavior-based safety incentives for their traditional safety incentive program that offers rewards for reductions in injury rate. If successful at this, the team's challenge changes to developing an acceptable and effective behavior-based safety incentive program. Then, the team needs to evaluate the impact of this incentive/reward program and refine it for another application. This plan-implement-evaluate-refine process needs to be repeated over the long term to maximize the beneficial impact of BBS incentives/rewards.

Earlier in Chapter 11, details on the design, administration, and evaluation of safety incentive/reward programs were presented. My point here is that each of the four key phases—planning, implementing, evaluating, and refining—implies a different team focus, along with unique goals and task assignments. Special training, resources, and individual talents are needed for each phase, requiring appropriate adjustment in team leadership, meeting agendas, and task assignments. These four phases are not peculiar to a Safety Incentive/Reward Team. They are relevant for each of the seven safety teams listed in Figure 17.6.

Nothing helps a team more to stay motivated and aligned with its mission than an objective presentation of their accomplishments, and an opportunity to learn how to improve their interventions for more achievement. This is the essence of an objective and equitable accountability system, which leads to people's responsibility for safety extending beyond the numbers. This is self-accountability or self-motivation (Geller, 2016; Geller & Geller, 2023), which is fundamental to attaining and maintaining an injury-free workplace.

Renewing. When team members observe the "fruits" of their labor, their motivation to continue their efforts is bolstered. In other words, observation of success breeds more success. Thus, an optimal way to renew the confidence and purpose of a team is to display clear and objective evidence that their efforts have made a positive difference. More can and should be done, however, to move teams forward with renewed concern and commitment.

A team-building session can be conducted with the sole purpose of restoring team members' motivation toward teamwork. Often, it is beneficial to hire an outside consultant or facilitator to conduct such a session. However, it is possible your company training department employs a person who could facilitate an effective team-building session.

17.2 IN SUMMARY

This is, obviously, only a brief overview of basic steps involved in developing and sustaining high-performance safety teams. Additional details related to each of these procedural steps are available in other texts (e.g., Geller & Geller, 2023; Lloyd, 1996; Parker, 1996; Rees, 1997). You realize, of course, that effective OSH teams do not develop overnight. Each process reviewed here takes time and patient application of the various interpersonal and group process strategies described in Parts 3 and 4 of this text, including behavior-based observation and feedback, proactive and empathic listening, directive and supportive coaching, individual recognition, and AC4P performance evaluations.

The benefits of implementing those teamwork strategies will not be immediate. The "sell" for teamwork is analogous to the "sell" for OSH. Safety leaders are well aware of the need to perform certain inconvenient, inefficient, and even uncomfortable safety-related behaviors in order to reap the potential long-term benefits of injury prevention. Likewise, the rewards of teamwork require substantial up-front investment in resources, time, and collective effort.

Many of us are not used to the collaborative and cooperative interdependency of high-performance teamwork. That includes the managers who hold us accountable for our work output. Therefore, the teamwork perspective of mutual accountability for shared goals needs to be appreciated by managers and supervisors as well as by team members. The people in organizations who provide the resources and opportunities for teamwork need to understand what it takes to reap the teamwork benefits of synergy, even if they are not members of a particular team themselves.

Figure 17.7 depicts a typical scenario in the hustle and bustle of our everyday lives. Everyone is doing his or her own thing from a win/lose, individualistic paradigm. The outcome seems like utter chaos and leaves individuals with the impression that their personal goals are temporarily

FIGURE 17.7 Systems thinking reflects the interdependency paradigm needed for high-performance teamwork.

thwarted. This can lead to frustration and a bad attitude toward the whole situation. A bad attitude can influence risky or win/lose behavior, which in turn adds fuel to a bad attitude.

Our hero in Figure 17.7 has a different perspective on the whole situation. He is able to take a broader view and appreciate the marvelous interdependent transportation system. This systems-thinking perspective toward everyday circumstances can be greatly beneficial to the safety and health of individuals and groups.

Systems thinking benefits teamwork, *and* it is a consequence of teamwork. Therefore, systems thinking fuels the interdependency and collectivistic mindset needed for high-performance teams, and productive teamwork fuels more systems thinking. This attitude-behavior spiral is constructive and motivates the special commitment and dedication needed to build and maintain successful OSH teams destined to achieve a TSC.

18 Evaluating for Continuous Improvement

Continuous improvement demands proper evaluation. This chapter explains how to evaluate the impact of OSH interventions from an environment, behavior, and person perspective. More employees need to contribute information pertinent to intervention evaluation. This chapter shows you how to make that happen. The principles described here will make you a smarter consumer of marketed safety programs, and help you evaluate your own customized intervention processes.

What gets measured gets done; what gets measured and rewarded gets done well.

(Larry Hansen)

Larry Hansen (1994) used those words in his *Professional Safety* article on managing occupational safety (page 41). You have probably heard words to that effect. Indeed, they are key to any continuous improvement effort. However, there is a problem with how OSH is traditionally measured. As indicated earlier in Chapter 3, too much weight is given to outcome numbers that people cannot control directly. People need to be held accountable for results they can control. Yet, corporations, divisions, plants, and departments are often ranked according to abstract outcome numbers like the organization's total recordable injury rate. Such rankings often determine bonus rewards or penalties.

What behavior improves when OSH awards are based only on an organization's injury rate? If employees can link their daily activities to safety results, then celebrating reduced injury rates can be useful, even motivating. It is critical, however, to recognize the behaviors, procedures, and processes that lead to fewer injuries or lower workers' compensation costs.

If you don't focus on the real causes of improvement, you run the risk of actually demotivating the folks who deserve recognition. Employees might think continuous improvement is caused by luck or chance—events beyond their personal control. This can lead to feelings of apathy or perceived helplessness, as discussed earlier (see Chapters 6, 15, and 16). If you want employees to work for continuous improvement, the "right stuff" needs to be recognized and rewarded. This requires the right kind of measurement procedures.

18.1 MEASURING THE RIGHT STUFF

Deming (1991) admonished his audiences for ranking people, departments, and organizations. In fact, he recommended that grades and performance appraisals be abolished completely from education and business. In his words, "The fact is that performance appraisal, management by the numbers, M.B.O., and work standards have already devastated Western industry . . . the annual rating of performance has devastated Western industry . . . Western management has for too long focused on the end product" (Deming, 1986a, p. 1).

Part of Deming's rationale comes from the fact that standard approaches to measuring academic and work performance are often subjective, relative, and not clearly related to individual behavior. Teachers and professors, for example, use contrived distribution curves and cut-offs to assure only a designated percentage of students can attain certain grades. Knowledge tests are necessarily biased and are imperfect assessment devices (especially multiple-choice exams), and they are often only remotely linked to specific behaviors within a student's control—including attending class, taking notes, reading the textbook, and studying the material on a regular basis.

I have known many demotivated students who felt their daily efforts were overshadowed by the emphasis on exam grades.

18.2 DEVELOPING A COMPREHENSIVE EVALUATION PROCESS

Years ago, I had the pleasure of working with a panel of evaluation experts to develop a set of measurement guidelines for the National Safety Council and the Centers for Disease Control. Our mission was to develop a handbook of practical guidelines that field personnel could use to evaluate the impact of an intervention process that was designed to improve safety on family and commercial farms.

We agreed that evaluation was essential to hold people accountable for achieving program objectives. This pertains particularly to those individuals developing and implementing the intervention. The evaluation process should measure whether intervention procedures: 1) are consistent with relevant principles and mission statements of the organization, 2) reach the desired audience and are implemented as planned, and 3) are efficient and effective. Thus, a complete evaluation process should assess how an intervention was conceived, designed, and implemented and how efficient and effective it was at improving safety-related behavior, person-states, and/or environmental conditions.

Figure 18.1 summarizes our panel's deliberations on how to evaluate intervention impact. It integrates the primary issues discussed in this chapter. Lower levels of the hierarchy represent process activities needed to improve the higher-level outcomes of a safer environment and to ultimately prevent injuries. The most immediate determinants of injury reduction are improvements in the environment or behavior or both the environment and behavior.

Let's pause for a moment to consider the cause-and-effect connection between process and outcome. Behavior can be viewed as an outcome, shaped by a process that pursues improvement in employee knowledge, perceptions, and/or attitudes. In this case, a behavior-change goal depends on employee participation in an intervention aimed at influencing one or more person-states. Of

FIGURE 18.1 Measures of intervention impact vary according to remoteness from immediate injury causation.

Evaluating for Continuous Improvement

course, intervention processes can be designed to circumvent person-states and directly change behavior for injury prevention, as discussed in Part 3 of this text.

Figure 18.1 illustrates the relativity of process and outcome. An education process, for example, can lead to behavior change (an outcome), while a process to improve behavior might result in outcome changes to the environment, like improved housekeeping. Completing a work process in a safer environment can affect the ultimate outcome—fewer injuries.

Figure 18.1 also reflects the three basic areas requiring attention for injury prevention—environment, behavior, and person. This is, of course, the Safety Triad introduced in Chapter 2 to categorize intervention strategies and referred to later in Chapter 14 to classify different types of AC4P behaviors. The hierarchical levels of intervention impact in Figure 18.1 reflect one or more of these three domains and suggest a particular approach to measurement. Changes in environments and behaviors can be assessed directly through systematic observation, but changes in knowledge, perceptions, beliefs, attitudes, and intentions are only accessible indirectly with survey techniques, usually questionnaires. Let's review these three basic domains of evaluation.

18.2.1 What to Measure?

Most safety interventions focus on either environmental conditions—including engineering controls—or human conditions, as reflected in employees' perceptions, attitudes, or behaviors. It might seem reasonable to evaluate improvement or change only in the domain that was targeted—environment, behavior, or person-state. When the target is corporate culture, employee perceptions or attitudes are typically evaluated. If behavior improvement is the focus, then behaviors are observed and analyzed in terms of their frequency, rate, duration, or percentage of occurrence, as reviewed in Chapter 8. When evaluating environments or engineering technologies, mechanical, electrical, chemical, or structural measurements are taken.

Culture-change consultants often advocate perception surveys in place of environmental audits. Presentations on behavior-based safety emphasize direct observation of work practices, often in lieu of the subjective evaluation of personal perceptions and attitudes. However, given the need for employees to "feel good" about a BBS process and the need for employees to participate continuously in managing and monitoring that process, it is critical to check perceptions and attitudes, along with ongoing behaviors. For example, a comprehensive evaluation of a simple change in equipment design should probably include an assessment of relevant human factors like employees' work behaviors around the new equipment, as well as their attitudes and perceptions regarding the equipment change.

Thus, a comprehensive evaluation for OSH requires a three-way audit process covering environmental conditions; safety-related behaviors; and related person-states such as perceptions, attitudes, beliefs, and intentions.

18.2.2 Evaluating Environmental Conditions

In many ways, environmental audits are the easiest and most acceptable type of evaluation. In fact, regular environmental or housekeeping audits are already standard practice at most companies. These evaluations can often be improved by involving more employees in designing the audit forms, conducting systematic and regular assessments, and posting the results in relevant work areas.

Figure 18.2 depicts a generic environmental checklist for OSH that can be used to graph for public display the percentage of safe conditions and the percentage of potential corrective actions taken for at-risk tools, equipment, or operating conditions. My colleagues at Safety Performance Solutions (SPS) typically taught work teams the rationale behind the environmental checklist and then assisted them at applying the checklist at their facility. Subsequently, employees customized the checklist and graphing procedures for their particular work areas. Regular audits

Observer: _____ Date: _____ Time: _____				
Department: _____ Building: _____ Floor: ____ Area: _____				
Operating Conditions/Tools & Equip.	**Safe**	**At-Risk**	***Corrective Actions Taken**	
Electrical wiring (properly enclosed)				
Air nozzles (limited to 30 P.S.I.)				
Chemicals (exposure concern)				
Eyewash station				
Emergency shower				
Barricades (in place where necessary)				
Storage of materials (neat/safe)				
Hazard Communication labels (appropriate)				
Floors (dry)				
Exits, aisles, sidewalks and walkways (clear of debris)				
Lighting (adequate)				
Housekeeping (satisfactory)				
Tools (safe operating condition)				
Guards (adequate and in place)				
Fire extinguisher (monthly inspection)				
Fire extinguisher (in appropriate location)				
GoJo (safe operation)				
GoJo driver (license in possession)				
Tow truck (safe operation)				
Tow truck driver (license in possession)				
Chairs in (safe condition)				
Totals				

Percent Safe Conditions: $\dfrac{\text{Total Safe Observations}}{\text{Total Safe Observations + At-Risk Observations}} \times 100 = $ ____ %

Percent Corrective Actions Taken: $\dfrac{\text{Total Corrective Actions Taken:}}{\text{Total At-Risk Observations}} \times 100 = $ ____ %

* Please list and define the corrective actions taken on back of this sheet.

FIGURE 18.2 An environmental checklist can be used to evaluate the safety of tools, equipment, and operating conditions.

and feedback sessions increase accountability for environmental factors that can be changed to prevent an injury.

The property-damage incident. A provocative book by Bird and Germain (1997) focuses on environmental assessment, in particular the need to investigate thoroughly the "property-damage accident." Then the property damage needs to be fixed in order to prevent a workplace injury. The authors claim the investigation and correction of property damage or equipment in need of repair are key to improving workplace safety. Yet the property-damage incident is sorely overlooked.

If you are underwhelmed by this so-called "missing link," I understand. I felt the same when Frank Bird first related his passionate thoughts about the property-damage incident. I did not appreciate the profound implications of this concept until reading his book and discussing it with line workers. For example, when I have introduced the property-damage incident at my workshops and seminars, line workers in attendance showed special interest. They often testified to the extreme amount of property damage at their work sites, including stockpiles of broken ladders; tools in disrepair; machine guards that do not work properly; and dents in equipment, walls, and vehicles. Each dent signals an incident, perhaps a close call, that was not analyzed.

Front-line workers have also verified the dramatic impact of property damage on their work demeanor, which in turn influences their attitude about OSH. To them, unrepaired property damage signifies that "Management doesn't care about our work situation," or "It's okay to damage property as long as we meet production demands."

An industrial example. Consider this true story that illustrates how a focus on property damage can make a difference. Walking along a scaffold, a worker slipped on a metal plate and almost fell several stories to his death. Fortunately, he was able to catch himself with his arms and pull himself back onto the walkway. Members of the safety committee decided to do more than the typical "reactive investigation" of that close call. They did not simply hold the welder responsible for not securing the plate. Instead, they looked for other contributing factors in order to prevent similar mishaps.

Guess what they discovered? At least a dozen people had slipped on that same loose plate and said nothing about it. No one reported a close call. They did not want to report an incident that could imply "careless or thoughtless behavior." However, if the loose plate had been reported as property damage that needed immediate repair, the idea of individual blame would have been removed. Thus, not reporting such property damage and assuring corrective action should be considered "careless and thoughtless."

Removing personal blame from incidents that set the stage for personal injury will enable more proactive reporting, evaluating, and correcting. When your periodic environmental audits show less and less property damage, you can be assured you are preventing injuries. In fact, this is actually a more reliable and valid metric for safety improvement than the standard injury and illness rates derived from employees' self-reports and visits to the plant infirmary. So, a comprehensive safety measurement system should include systematic audits of damage to the work environment. The repair of environmental damage should be continuously tracked as an ongoing measure of OSH improvement.

18.2.3 EVALUATING WORK PRACTICES

The systematic auditing of work practices was the theme of Chapters 8, 9, and 12. In Chapter 8, the overall DO IT process was introduced—"D" for define target behaviors, "O" for observe target behaviors, "I" for intervene to increase occurrences of safe behavior or decrease occurrences of at-risk behavior, and "T" for test (or evaluate) the impact of your intervention.

Strategies for developing two types of behavioral observation checklists were discussed in Chapter 8—a generic version for basic work practices applicable anywhere, such as prescribed lifting techniques and the use of certain PPE (see Figures 8.6 and 8.7), and a job-specific checklist for particular tasks, like the safe-driving checklist my daughter and I used (see Figure 8.5).

The coaching process detailed in Chapter 12 also discussed how to develop and apply both generic and job-specific checklists for one-on-one observation-and-feedback coaching sessions with coworkers. As mentioned previously, using a behavioral checklist to observe and evaluate ongoing work practices is the type of performance appraisal that can lead to continuous improvement.

Chapter 9, "Behavioral Safety Analysis," addressed evaluation from the perspective of work practices. A series of ten questions was proposed for conducting a step-by-step examination of the situational, social, and personal factors influencing at-risk behavior. Answers to those questions (see Figure 9.4) provide direction for deriving a most cost-effective plan for corrective action.

18.2.4 EVALUATING PERSON FACTORS

As discussed earlier in this text, person factors refer to subjective or internal aspects of people. They are reflected in commonly used terms like attitude, perception, feeling, intention, values, intelligence, cognitive style, and personality traits. You can find many surveys that measure specific person factors of target populations ranging from children to adults. Some of these factors are presumed to be traits; others are considered states. It is important to understand the difference when you consider the evaluation potential of a particular survey.

Person-traits. Theoretically, traits are relatively permanent characteristics of people that do not vary much over time or across situations. Because traits are relatively permanent, questionnaires that measure them cannot gauge the impact or progress of an OSH intervention. Trait measures serve as a tool to teach individual differences, but for OSH, their application is limited to selecting people for certain job assignments. However, that approach is not very effective, as discussed in Chapter 1.

Person-states. Person-states are human factors that can change from moment to moment, depending on situations and personal interactions (as discussed in Chapter 15). When our goals are thwarted, for example, we can experience a state of frustration. When experiences lead us to believe we have little control over events around us, we can be in a state of apathy or helplessness. Person-states can certainly influence behavior. Frustration, for example, often provokes aggressive behavior, and a perception of helplessness inhibits constructive behavior and facilitates inactivity.

In contrast, certain life experiences can affect positive person-states, such as optimism, self-efficacy, self-confidence, and belongingness. These, in turn, boost constructive behavior. This is the indirect approach to increasing AC4P behavior discussed in Chapter 16. The woman in Figure 18.3 is in a positive person-state referred to as optimism. She might drive her friends crazy, but research has shown that healthier and happier people are more often in this state. Plus, as discussed earlier in Chapter 15, optimistic people are more likely to actively care for the safety and/or the health of others.

Measures of person-states can be used to evaluate perceptions of OSH and to pinpoint areas of a culture that need special intervention attention. Like most culture surveys, our Safety Culture Survey asks participants to answer questions on a five-point continuum (from highly disagree to highly agree) about their perceptions of safety-related issues. Issues include the perceived amount of management support for safety, the willingness of employees to correct at-risk situations and

FIGURE 18.3 An optimistic person state facilitates happiness, perseverance, achievement, health, and actively caring.

Evaluating for Continuous Improvement

look out for the safety of coworkers, the perceived risk level of the participant's job, and the nature of interpersonal consequences following an injury.

Our culture survey also measures factors that increase one's willingness to actively care for another person's safety. These include self-esteem, belongingness, and empowerment, as detailed in Chapter 15. Figure 18.4 contains 20 items from the safety-perception-and-attitude portion of our culture survey. You will note nothing very special about the items in that scale. They ask employees to react to straightforward statements about safety management and improvement.

You could compare employees' reactions to the items in Figure 18.4 before and after implementing a safety improvement process. Studying reactions prior to an intervention helps identify issues or work areas needing special attention. Such information can lead to choosing a particular intervention approach or to customizing an OSH intervention. Data from a baseline perception survey might even indicate that a culture is not ready for a given intervention process, suggesting the need for more education and discussion to get employees to "buy in."

	Item	Highly Disagree	Disagree	Not Sure	Agree	Highly Agree
1.	The risk level of my job concerns me quite a bit.	1	2	3	4	5
2.	When told about safety hazards, supervisors are appreciative and try to correct them quickly.	1	2	3	4	5
3.	My immediate supervisor is well informed about relevant safety issues.	1	2	3	4	5
4.	It is the responsibility of each employee to seek out opportunities to prevent injury.	1	2	3	4	5
5.	At my plant, work productivity and quality usually have a higher priority than work safety.	1	2	3	4	5
6.	The managers in my plant really care about safety and try to reduce risk levels as much as possible.	1	2	3	4	5
7.	When I see a potential safety hazard (e.g., oil spill), I am willing to correct it myself if possible.	1	2	3	4	5
8.	Management places most of the blame for an accident on the injured employee.	1	2	3	4	5
9.	"Near misses" are consistently reported and investigated at our plant.	1	2	3	4	5
10.	I am willing to warn my coworkers about working unsafely.	1	2	3	4	5
11.	Employees seen behaving unsafely in my department are usually given corrective feedback by their coworkers.	1	2	3	4	5
12.	Compared to other plants, I think mine is rather risky.	1	2	3	4	5
13.	Working safely is the number one priority in my plant.	1	2	3	4	5
14.	I have received adequate job safety training.	1	2	3	4	5
15.	Many first aid cases in my plant go unreported.	1	2	3	4	5
16.	Information needed to work safely is made available to all employees.	1	2	3	4	5
17.	Management here seems genuinely interested in reducing injury rates.	1	2	3	4	5
18.	Safety audits are conducted regularly in my department to check the use of personal protective equipment.	1	2	3	4	5
19.	I know how to do my job safely.	1	2	3	4	5
20.	Most employees in my group would not feel comfortable if their work practices were observed and recorded by a coworker.	1	2	3	4	5

FIGURE 18.4 These questionnaire items measure personal perception regarding the safety of an organization (selected from the Safety Culture Survey developed by Safety Performance Solutions, Inc.).

18.2.5 Evaluating Costs and Benefits

A comprehensive cost–benefit analysis is invaluable in sustaining top-management support for a safety-improvement process. It also provides motivating feedback for the program participants. Thus, it is important to maintain records of direct and indirect costs associated with injuries—and with injury prevention—even if those calculations are only estimates. Comparing estimates with the costs of implementing and maintaining particular OSH programs illustrates specific benefits and justifies continued program support, especially if you can show that a particular intervention process has substantially reduced injury frequency and costs.

When calculating program costs, you should document every expense, including promotional materials, teaching aids, evaluation supplies, rewards, media expenditures, and wages paid for intervention assistance. Employees' time away from the job to plan, present, evaluate, or participate in the intervention process should be estimated, even if the time is voluntary—for example, on evenings or weekends. If you are comprehensive when calculating program costs, then you are justified in estimating the numerous direct and indirect costs resulting from a job-related injury.

Injury records should be consulted before and after an intervention process has been started in order to show the savings from fewer work-related injuries. Direct costs that should be calculated per injury include:

- Wages paid to absent employees (workers' compensation).
- Property damage.
- Medical expenses.
- Physical and vocational rehabilitation costs.
- Survivor benefits.

Those direct costs may be the proverbial tip of the iceberg when considering the indirect or hidden costs of business disruptions caused by a lost-time injury. However, indirect costs can be difficult or impossible to calculate. You should, however, try to estimate such costs in these categories:

- Overtime pay used to cover the work of an injured employee.
- Scheduling work tasks to cover for an injured employee.
- Additional administrative hassles, extra wages, training time, and inefficient work associated with temporary replacements.
- Special costs of losing a skilled employee.
- Extra time from work supervisors to schedule shift changes, temporary replacements, or employee training necessitated by the absence of an injured worker.
- Retraining and readjusting for employees returning to work after an extended absence. (A permanent disability from an injury might call for a new job assignment.)
- Special costs for extensive recruitment procedures and on-the-job training for permanent replacements for injured employees who do not return to work or who return with a permanent disability.
- Special administrative costs to investigate and document the incident and the medical treatments for compliance with state workers' compensation laws and with other state and federal regulations, such as OSHA standards.

Consider that the direct and indirect costs to a company from a work-related injury can be overwhelming. It is certainly an intimidating chore to estimate these costs, and it is stunning to consider the corporate losses from just one employee's injury. Those dollars clearly justify considerable OSH intervention design, implementation, and evaluation. Just anticipating the negative consequences from one work-related injury should motivate support and participation in proactive OSH efforts. This is the first critical step in "selling" safety—the theme of the next chapter.

18.3 YOU CAN'T MEASURE EVERYTHING

Deming (1991, 1992) condemned grades and performance appraisals because they provide a limited picture of an individual's contributions and potential. They might also constrain the number and the type of interventions used to improve the quality of a work culture. If, for example, the only procedures implemented to improve OSH are those that enable objective measurement, the number and quality of safety interventions would be severely restricted.

Chapter 16 introduced a number of ways to increase occurrences of AC4P behavior—directly through applications of learning and social influence principles and indirectly through improving the person-states that increase one's willingness to actively care. However, it is impractical and impossible to measure the impact of many of those interventions. Should we avoid creating and implementing OSH interventions that cannot be objectively evaluated?

Dr. Deming explained that there are many things we should do for continuous improvement without attempting to measure their impact. We should not do such things only to influence performance indicators but because they are the right things to do for people. You might never be able to measure the impact of treating an employee with special respect and dignity, but you do it anyway. Such treatment may, in fact, contribute to achieving a TSC, but you will never know. Likewise, you will never know how many injuries you prevent with proactive AC4P behavior, and you will never know how much AC4P behavior you will promote by taking even small steps to increase coworkers' self-esteem, empowerment, and sense of belonging. You need to continue doing those things anyway. Many things that cannot be measured and rewarded still need to get done.

18.4 IN SUMMARY

At the start of this book, I explained the fallacy of basing decisions on commonsense. Rather than adopt intervention programs that sound good, we need to use procedures that work. But how do we know what works? Of course, you know the answer to that question. Only through rigorous program evaluation can we know whether an intervention is worth pursuing. Now comes the more difficult question. What kind of program evaluation is most appropriate for a particular situation?

Actually, every chapter of this text has addressed program evaluation in one way or another. Early on, I explained the need for achievement-oriented methods to keep score of your safety efforts. That enables people to consider safety in the same work-to-achieve context as production quantity and quality. That implies, of course, the need to use program-evaluation numbers people can understand and learn from. This is how evaluation leads to continuous improvement.

The intervention information presented in this text is founded on rigorous evaluation, not commonsense. Evaluation techniques used in published research are, indeed, more rigorous and complex in terms of reliability, validity, and statistical analysis than those needed for continuous improvement of real-world OSH programs. The basic principles and issues presented in this chapter, however, are relevant to both researchers (seeking to contribute to professional scholarship) and practitioners (seeking continuous improvement of an intervention process).

To publish their findings, researchers need to demonstrate reliability and validity of their measures and find statistical significance. However, some researchers can and do ignore several evaluation principles presented in this chapter. For example, their measures typically target only one dimension (environment, behavior, or person factor), are short-term (applied for a limited number of observation sessions), are subjected to statistical transformations and analyses that take substantial time to complete and are not readily understood by the average person, and often do not include a cost–benefit analysis. Reports of their procedures and results only need to be understood and appreciated by a select, often esoteric, group of professionals who specialize in the particular issue or problem addressed by the research.

However, you should not overlook the basic principles presented in this chapter when evaluating practical interventions to achieve continuous improvement in OSH. Data-collection procedures and statistical analyses can often be less rigorous, but safety practitioners need to address several important issues that are often bypassed by professional researchers. Specifically, OSH practitioners need to:

1. Define the level of performance targeted by the intervention while appreciating limitations in targeting individual vs. organizational performance.
2. Use procedures to assess the three dimensions of safety improvement—environment, behavior, and person factors.
3. Apply process measures periodically over the long term, especially checks on environmental conditions and work practices.
4. Include a cost–benefit analysis to justify continued intervention and evaluation efforts.
5. Keep score with numbers that are both meaningful to all program participants and provide direction for intervention refinement.

Those last two guidelines are critical to meeting the challenge addressed in the next chapter—obtaining and maintaining support for an effective intervention process.

19 Obtaining and Maintaining Engagement

You cannot effectively apply the principles in this book without ongoing support from both managers and employees. This chapter focuses on ways to initiate and maintain that support, including ways to promote leadership, build commitment and engagement, expand the scope of interventions, reduce active resistance, and sustain momentum.

Character consists of what you do on the third and fourth tries.

(James Michener)

Culture change is never quick and never easy. The "quick-fix" illustrated in Figure 19.1 is clearly ridiculous. As absurd as that notion is, it comes to us naturally. We want speedy solutions to difficult challenges. It is easy to lose patience, enthusiasm, and optimism along the way. After all, our society demands immediate gratification—just look at all the movies and television shows that begin with dramatic problems and come to happy endings within 30 to 90 minutes. Plus, the faster we solve any problem, the sooner we experience positive rewarding consequences.

This chapter brings us to the point of pulling things together. Let's consider the broad challenge of initiating a culture-improvement process aimed at achieving a TSC. First, some general guidelines are provided for starting a process and maintaining support. Then, the critical concepts of leadership, communication, and resistance are addressed. You will not find step-by-step cookbook procedures here; a generic recipe is just not available. Instead, take the principles and procedures presented in prior chapters, add the information presented here, and you will be well on your way to an innovative experience in improving OSH and achieving a TSC.

FIGURE 19.1 There's no quick-fix solution to culture change.

19.1 STARTING THE PROCESS

19.1.1 Management Support

You cannot progress without support from management. How many times have you heard, "Whatever management really pushes and supports will happen"? Implicitly, "Whatever upper management does not push and does not support will fail." If managers emphasize housekeeping, quality control, or cost-reduction, improvements in those domains are likely to follow. Strong top-down support, involvement, and commitment alone will not make a campaign succeed, but they are essential ingredients. Plus, management and labor must collaborate to make any process work.

19.1.2 Creating a Safety Steering Team

A Safety Steering Team plays a critical role in developing a TSC, providing policy-making, oversight, and general support. All of this is simply more than any one person can handle. However, before creating a Safety Steering Team, it is important to examine the existing committee structure in your organization. You don't want to duplicate the efforts of the safety department or some other relevant standing committee. For example, a current employee team might be able to take on the responsibility of coordinating efforts to cultivate and achieve a TSC.

Careful planning is needed to determine:

- What is the mission of the Safety Steering Team?
- What are the ground rules for how it operates? (see Chapter 17)
- What are the group's limitations or restrictions?
- What are the priorities?
- Who should be on this team?

19.1.3 Developing Evaluation Procedures

"Did it do any good?" This is the central question the Safety Steering Team must be prepared to answer about any intervention. Chapter 18 offered guidelines for evaluating intervention impact. At the start of an evaluation process, these questions need to be answered.

- What indicators should we examine—behaviors, attitudes, opinions, or outcomes?
- When should we evaluate?
- What types of data should we collect and analyze?
- What is the cost of this evaluation process?
- How should we summarize and display results?

19.1.4 Initiating an Education and Training Process

Ensuring that employees learn key principles and procedures to improve OSH is a primary responsibility for the Safety Steering Team. At the minimum, the following components should be incorporated into planning an effective education and training program:

- Develop education content and procedures.
- Plan the education and training process.
- Plan for follow-up sessions.
- Identify and prepare instructors.
- Measure the impact of the education/training process.

Obtaining and Maintaining Engagement

Let's discuss each of these elements in more detail, keeping in mind that these suggestions need to be customized at the plant level for optimal success.

Identifying and preparing instructors. Selecting the "right" instructors is critical because teaching is at the heart of effective education and training. If the teachers do a poor job, they undermine other training tools such as videos and booklets. You should consider these factors when choosing instructors.

- Prior experience in educating and/or training.
- Current level of teaching ability (aptitude and achievement).
- Credibility with employees to be educated.
- Level of motivation and interest in doing the instruction.
- Prior familiarity with psychology, especially applied behavioral science.
- Belief that the principles and procedures can help achieve a TSC.

To help selected teachers prepare for their task, they will need to: 1) understand the principles and relevant procedures in this text so they can represent them accurately to the participants; 2) feel comfortable with the specific process of the plant-wide education and training, which, ideally, they will help to develop; 3) practice basic communication skills, 4) demonstrate leadership at meetings; and 5) learn how to facilitate discussions—particularly if they have little experience at behavioral training and small-group meetings.

An in-house training staff can help volunteers build their teaching skills. These topics are often covered in various corporate training programs. Plus, instructional programs are readily available from outside vendors or consultants.

Developing course content. When considering the specific topics to include in each instructional session, you need to address these points.

- What are the specific goals of a particular session—for example, to teach principles, train procedures, or build commitment and motivation?
- In addition to this text, what sources are available for relevant content and support, such as local case studies of behavior-based safety?
- What relevant films, videotapes, and other instructional materials are available?
- To meet session goals, what key content points should be covered, and in what order?
- What specific facts, statistics, and case studies from your organization can be incorporated into the sessions?
- How much information can be covered effectively in one session?

Involve your selected instructors as much as possible in developing the specific education and training plan and process.

Planning the instructional process. "Everything was covered but nobody paid much attention." This is a common complaint about education or training. The translation is that good content is important but not sufficient. You need to "package" your content and present it in a way that hits home with the participants. Obviously, you want them to practice what is preached. Most instructors know their material; the challenge lies in conveying that knowledge. How do you get your message across effectively? Here are some points to consider:

- An interactive/participative approach is typically more effective than a "top-down" lecture coming from the podium.
- Like a good pitcher, change speeds. Do not rely on one pitch or one way of presenting information.
- It is easier to involve small groups of participants than large ones.

- Regardless of the main objective of a session, some initial awareness-raising makes participants more receptive to the content.
- Integrate demonstrations into the program.
- If possible, have participants practice a particular skill taught with appropriate feedback in the classroom or on the job.
- Resolve those administrative questions, including where and when the instructional sessions should be held and how long they should take.

Evaluating effectiveness. As covered in Chapter 18, you cannot overlook evaluation. At this point in the overall intervention process, you need to determine the impact of the education and training on participants' knowledge, skill, and attitude.

Knowledge of content can be assessed the old-fashioned way, through written tests given at the end of a session. Skills can be evaluated by systematic behavioral observation in the classroom or in the workplace. Participants' reactions to the session can be measured with brief questionnaires. For this you should consider the following:

- Brevity.
- Relevance and complexity of wording.
- Combining objective ratings and written comments.

19.1.5 SUSTAINING THE PROCESS

Continued upper-management support. Many safety programs get a big send-off, only to drop off the radar screen. Then, it is "out of sight, out of mind" as some new program is pushed. This is why safety is often derided for its "flavor of the month" approach. So, how can you sustain an OSH process? (Note the term "process" rather than "program" here, because processes flow on, while programs begin and end.) First and foremost, you need continued and visible support from top management. If management endorses the process on an ongoing basis, it can become integrated into normal plant operations.

The Safety Steering Team needs to work diligently to convince managers that their commitment is fundamental to the success of an OSH process—not only to get it going but to keep it going. Here are some thoughts on maintaining that all-important support of top managers:

- First, you need to gain access to upper management. Identify a manager to champion your cause in the executive offices. Bring him/her into the loop, ask him/her to attend all team meetings.
- Keep managers informed. Submit team reports to them on a regular basis but do not overwhelm busy managers with minutiae.
- Keep managers involved. Solicit their comments and "concerns" about the process you have underway. Of course, you should be doing this at all levels of the organization to create a top-to-bottom sense of ownership.
- Promote and market your efforts. Publish articles and announcements about the OSH process on the company website and in the employee newsletter on a regular basis.
- Keep at it. Identify the benefits of your process and continue to "sell" them to upper management.

19.1.6 FOLLOW-UP INSTRUCTION/BOOSTER SESSIONS

Even with ongoing support, a comprehensive OSH-improvement process cannot succeed without carefully planned follow-up instruction. From time to time, education and training content must

Obtaining and Maintaining Engagement

be updated to reflect changes in plant conditions, the use of new machines or protective equipment, and the like. Do not delay in keeping pace with change; what is being taught should correspond exactly to current plant conditions. Consider the following issues/questions for follow-up education and training:

- When and how should basic instruction be repeated for new employees?
- What are the objectives of follow-up instruction?
- How should new material be integrated?
- How often should follow-up sessions occur?
- What should be the content?
- How should the material be presented—film, lecture, on-the-job training, discussion groups, audio tapes, interactive computer program?
- How can monthly safety talks on team meetings be used as boosters or activators?

19.1.7 Troubleshooting and Fine-Tuning

Once a safety achievement process is up and running, the Safety Steering Team confronts the responsibility of fine-tuning the procedures. This is based on ongoing evaluations. If the process is going to be sustained, employees and managers must perceive it as current—state of the company—in terms of relevant, adaptable, and responsive content. It simply cannot be "frozen" nor left unattended. You cannot "wind up" a process at the start and expect it to run forever like that battery-powered bunny. Some keys to fine-tuning include:

- Discuss the impact—is it working? What do the data from participants' reaction sheets, as well as other measures, tell you?
- Identify strengths and weaknesses. Based on the data, which elements should be kept, changed (and how), or replaced?
- Cope with change. Be sure all those affected by the process are fully informed about changes when they occur. Ideally, all participants should be actively involved in troubleshooting and fine-tuning the OSH interventions.

19.2 CULTIVATING CONTINUOUS SUPPORT

Starting a safety improvement process and maintaining it over the long term requires the three essential support processes depicted in Figure 19.2. Communication and recognition were covered in Chapter 13. Here, the focus is on leadership. Leaders are needed to champion new principles and procedures. In fact, leadership makes the difference between a "flavor of the month" OSH initiative and a long-term continuous improvement process.

My colleagues and I at Safety Performance Solutions have seen the principles and procedures presented in this book lead to remarkable success and eventually to a TSC. All too often, however, we have seen good intentions and superb introductory instruction fizzle out and go nowhere. Why? It is a function of leadership. You can launch a process with excellent education and training, but you cannot keep the momentum going without individuals who provide energy, enthusiasm, and the right example. This section covers some essentials of effective leadership.

19.2.1 Where Are the Safety Leaders?

First, we have to find the leaders. Who are they? The traditional definition of one person exerting influence over a group does not work for OSH. Ask any safety manager who has been expected

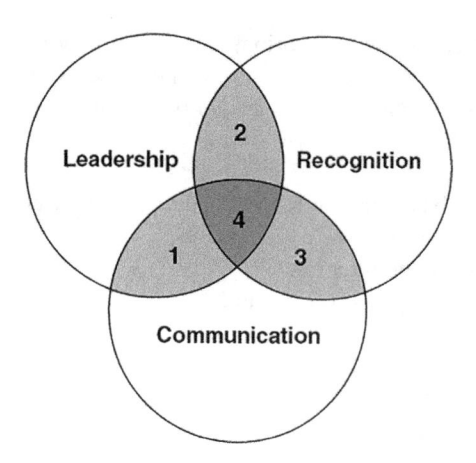

1. Leaders *communicate* effectively.
2. Leaders *recognize* desired performance.
3. *Recognition* is *communicated* effectively.
4. Leaders *recognize* desired performance effectively through a variety of communication channels.

FIGURE 19.2 Continuous improvement depends on three support systems.

to do it all. To achieve a TSC, everyone needs to accept a leadership role in preventing injuries. Everyone needs to feel responsible for safety and go beyond the call of duty to protect others. This requires leadership skills, including the effective delivery of supportive feedback for another person's safe behavior and corrective feedback for at-risk behavior.

Psychologists have studied leadership rigorously for over 70 years in an attempt to define the qualities of effective leaders. Still, many questions remain unanswered, making leadership more an art than a science. Several decades of research, however, have revealed some useful answers that can be applied to OSH.

Passion. The most successful leaders show energy, desire, passion, enthusiasm, and constant ambition to achieve. Passion to achieve a TSC can be fueled by clarifying goals and tracking progress. Put a positive spin on safety, make it something to be achieved—not losses to avoid. Then, employees will be motivated to achieve shared safety goals just like they work toward reaching production quantity and quality goals. Recognizing and sharing the achievement of OSH goals leads to the genuine belief that the process works. This motivates employees to continue the process.

Honesty and integrity. Effective leaders are open and trustworthy. A TSC depends on open interpersonal conversation. This obviously requires honesty, integrity, and trust. It is quite useful for work groups to discuss ways to nurture these qualities in their culture. Take a look at certain environmental conditions, policies, and behaviors. Some might arouse suspicions of hidden agendas, politics, and self-serving aims. You can work to eliminate some of those trust-busters by first identifying them, discussing their purpose, and devising alternatives. While frankness is important for increasing trust, it is important to be tactful when communicating an honest opinion, as not shown in Figure 19.3.

Motivation. Because most people naturally care about reducing personal injuries, even to people they do not know, the motivation to lead others for OSH will spread naturally throughout a work culture when people believe they can have personal control over injuries. This occurs when they learn effective techniques to prevent injuries (as presented in Parts 3 and 4 of this text) and feel empowered to apply them (as covered in Part 5).

Self-confidence. Effective leaders trust in their own abilities to achieve. Education helps convince people they can achieve, but they need ongoing support and recognition for their efforts.

Obtaining and Maintaining Engagement

FIGURE 19.3 Candor should be delivered tactfully.

For example, the self-confidence needed to give safety-related feedback can be initiated with appropriate education and training, and can be maintained with interpersonal coaching, communication, and recognition. Notice that enhancing the three empowerment beliefs—self-efficacy, response-efficacy, and outcome-expectancy—as discussed in Chapter 16, builds self-confidence.

Thinking skills. Successful leaders can integrate large amounts of information, interpret it objectively and coherently, and act decisively as a result. Constructive thinking skills evolve among team members when objective data are collected on the progress of OSH interventions and used to refine or expand those processes and to develop new ones.

When teams work through the DO IT process (as covered in Part 4), participants develop skills to evaluate behavioral data and use that information to make intervention decisions. This is basic scientific thinking, the key to substituting profound knowledge for commonsense. Such mindful learning (Langer, 1997) and critical thinking lead to special expertise.

Expertise. To achieve a TSC, everyone needs to understand the principles behind the policies, rules, and interventions implemented to improve OSH. When employees teach those principles to coworkers, they develop the level of profound knowledge, expertise, and responsibility needed for exemplary leadership.

Flexibility. Successful leaders size up a situation and adjust their style accordingly. At times, some groups and circumstances call for firm direction—an autocratic style. At other times, the same people might work better under a nondirective, hands-off approach—a democratic style. The best leaders are good at assessing people and situations, and then matching their behavior to fit the need.

19.3 OVERCOMING RESISTANCE TO CHANGE

"How do we deal with people who resist change?"
"How do we get more people to participate?"

I have frequently received those questions at education seminars and workshops. First, let's face reality. Change is unpleasant for most people, and some are apt to react poorly. Change can threaten our "comfort zones"—those predictable daily routines we like to control. In fact, it takes a certain amount of personal security and emotional intelligence to try something new. A certain kind of risk-taking is needed to lead change, and some people want no part of exploring the unknown.

We have all been in unfamiliar situations where we are not sure how to act. We feel awkward and uncomfortable. If someone gives us direction that helps increase our sense of control, it is easier to adjust. We might even help others deal with the change. However, without leaders and adequate tools to cope with change, we might retreat, withdraw from the situation, or even actively resist the change.

So how do we deal with resistance? Simply put, we should teach people the skills and give them the tools to handle change, plus support those who set the right examples. This seems logical and intuitive, but it does not always happen. Instead, managers too often try to identify the malcontents and discipline them for not participating.

Let's try to understand resistance by considering one of the classic awkward and strange situations thrust upon many of us—our first school dance. Remember it? For me, it was a high school homecoming dance in 1957.

Remember your first dance. Attending your first dance is like a rite of passage. If you were anything like me, you were a bit nervous about this change in your social world. You might have been prepared for it. Family, friends, and teachers probably told you what to expect. Maybe, you even had dance lessons, but those "tools" did not make it any easier for some of us to participate.

Not for me, anyway. I did not participate, but I wanted to. Before the dance, I practiced how to ask a girl to dance. I took four, two-hour dance lessons at an Arthur Murray Dance Studio. I felt ready, but never once did I dance that night. I didn't feel too embarrassed, though, because there were so many others not dancing. As was the custom, boys stood on one side of the gym and the girls on the other.

As illustrated in Figure 19.4, some kids were dancing and seemed to be having a great time. They danced almost every number and tried to lure others out on the floor. I could not be enticed, though. I hung back in my comfort zone, but at least I was in the dance hall.

As illustrated in Figure 19.5, some students stood around in the parking lot, talking, drinking, and smoking cigarettes. Those were the resisters. Some were *active* resisters. They stayed in their

FIGURE 19.4 Different reactions to change can be seen at the high-school dance.

FIGURE 19.5 Some actively resist change while others follow.

cars, never intending to enter the dance hall. Now and then, those guys started up their cars and cruised around town for a while, and then returned to the parking lot. They would persuade others to hop in their car, try some beer, smoke a cigarette, fool around, or cruise.

Levels of participation. There are essentially five ways of reacting to change—call them levels of participation—and they were all on display at that dance. First, there are the true leaders who get totally involved. They are the innovators—those who view change as necessary and an opportunity to improve. At the dance, they were the teenagers on the floor for almost every number. They had the most fun. They did not necessarily know what they were doing when it came to dancing, but they got out there and tried. They took a risk. They got totally involved and benefited most from the occasion. A dance might start with only a few of those "risk takers," but those leaders often persuaded a number of others to get involved as the night wore on.

Some people want to change but need direction and support. They are motivated to participate but need models or leaders. At the dance, those were the kids who hung back at first. With a little encouragement, they danced a few numbers. By the end of the evening, you could not get them off the dance floor. They were now totally engaged.

Most of us are at the third level of participation. We are ready to get involved, but we will stay in our comfort zones until we are directed and motivated to participate. It might look like we are resisting change, but not really. Call us neutral when it comes to our attitude about change. We just are not sure what to do. We need self-confidence that we can handle the change. We also need genuine support (positive recognition) when we try to participate. Once in a while, just getting started, or "breaking the ice," is enough to turn a passive observer into an active participant. For the most part, however, we stand on the sidelines and watch. This level of participation was represented by the boys and girls who lined each side of the gym.

Types of resistance. The final two levels of participation are passive and active resistance. Passive resisters perceive change as a problem. They complain a lot. They are critical and untrusting of something new imposed on them. They seem to see only the negative side of a new program, policy, or challenge. They rationalize their position by gathering with others at their same level of nonparticipation, and they grumble and whine about proposed changes or about others who are participating in a change effort. Their whining and complaining usually stops when participation in a new process is clearly appreciated by the majority. Passive resisters are followers, and they will do what they see most people doing.

Those are the teenagers who came to the dance because everyone else would be there, but they felt so insecure or anxious that they did not enter the building. They looked for others hanging around outside and made fun of the silly dancing going on inside. Sometimes, those nonparticipants ran into an *active* resister.

Fortunately, active resisters are few in number, but it doesn't take many of these characters to slow down a change process. These individuals view change as a personal threat and/or an opportunity to resist. They see any change effort that was not their idea as a potential loss of personal control, and they often exert countercontrol to assert their control or freedom. Many parents observe this phenomenon when their children reach the teens. Most teenagers want to feel independent and, at times, will disobey their parents' directions—break the rules—to gain a sense of independence or self-control.

We all feel overly controlled at times and, perhaps, react to regain independence or assert personal freedom. Sometimes, our reactions are not thoughtful, caring, or safe. Active resisters feel the need to resist change, the status quo, or authority much of the time. This is partly because their contrary behavior brings them special attention—peer recognition for resisting.

Who gets the attention? Active resisters stand out and attract attention. Nonparticipants use them to rationalize their own commitment to stay in their comfort zone. Managers monitoring the workplace often hit resisters with discipline, but this can backfire. This makes the top-down control more obvious for resisters. Such "discipline" builds their resentment of the system and makes it even less likely they will join the change process.

For some individuals, disciplinary attention only fuels their burning desire to exert independence and resist change. As a result, they might become more vigorous in recruiting others to oppose change. As discussed in Chapter 11, top-down discipline (or negative consequences) should be used sparingly if the ultimate purpose is total participation in an improvement process.

How were resistant teenagers brought inside to the dance? The harsh warnings of the school principal shouting from the steps did not work; neither did the one-on-one confrontation between one of the adult chaperones and the "leader of the pack." Whenever I saw a resister come inside, it was always the result of urging by another teenager. Peer pressure (or peer support) is still the most powerful motivator of human behavior.

Therefore, the best way to deal with resistance is usually to arrange for situations that enable or facilitate peer influence. Eventually, some of the resistant teenagers came into the building so they would not miss something. As depicted in Figure 19.6, they saw from their remote comfort zones that the people inside were really enjoying themselves, and they may eventually choose to participate.

The dance party will become more enticing as more and more teenagers dance. To increase such active involvement, the right kind of encouragement and support is needed. Will motivational lectures from a teacher, counselor, concerned parent, or outside consultant make that happen? It might make a temporary difference but not over the long term. The best way to deal with nonparticipation is usually to set up situations that allow for peer influence. This could mean managers do nothing more than support the change process and let peer pressure or support occur naturally.

FIGURE 19.6 When a critical mass of the culture changes, others follow.

19.4 PSYCHOLOGICAL SAFETY

A human dynamic that determines the level of employee engagement in OSH, as well as the amount of resistance to change, has been receiving significant attention in the corporate world these days. I am referring to "psychological safety"—a culture in which participants feel accepted and respected, and believe they will not be humiliated or reprimanded when speaking up with questions, concerns, mistakes, or new ideas (Geller, 2022).

Timothy Clark (2020a) describes four stages of psychological safety: 1) inclusion safety, 2) learner safety, 3) contributor safety, and 4) challenger safety. Inclusion safety is the foundation stage when individuals feel accepted in their group or work team. They sense a comfortable person-state of belongingness and interdependency, as discussed in Chapter 15.

In the next stage—learner safety—individuals participate in ongoing learning and teaching. As learners, they openly demonstrate vulnerability by asking questions and requesting help without fear of ridicule or embarrassment. They solicit feedback to be the best they can be, and whenever appropriate, they provide feedback to teach others. In other words, team members at this stage of psychological safety cultivate ongoing continuous improvement as both learners and teachers, depending upon situations and circumstances.

When employees or team members give others supportive and/or corrective feedback to improve behavior, they have reached the stage of "contributor safety." With a make-a-difference mindset, these individuals feel empowered to make meaningful contributions to their team and the organization. Then, participants realize the ultimate level of psychological safety when they feel comfortable challenging the status quo. Dr. Clark refers to this person-state as "intellectual bravery"—a "willingness to disagree, dissent, or challenge the status quo in a setting of social risk in which you could be embarrassed, marginalized, or punished in some way" (2020b, p. 10).

19.4.1 Relevance for OSH

As indicated earlier, the injury-prevention success of behavior-based safety depends on the delivery and acceptance of one-to-one behavior-focused feedback. After completing an employee-derived CBC, including reports of environmental factors that might have influenced the observed behavior, the observer communicates the results to the observed worker. Unfortunately, many organizations and their employees bypass this crucial BBS step, and deliver the checklist information to an individual who enters the data in a computer program that computes averages for comparing safe and at-risk percentages across work teams, both within and between organizations.

Why do many BBS applications omit the critical interpersonal feedback process? An obvious answer: a lack of psychological safety. What does that mean in practical operational terms? What is deficient in an organization that omits the invaluable interpersonal conversation component of BBS? Yes, we are back to considering the organizational culture—the environmental and social context of the workplace that can facilitate or stifle psychological safety.

19.4.2 Cultivating Psychological Safety

Given the obvious benefits of psychological safety, we arrive at the pivotal question: How can an organization cultivate a work culture that promotes, supports, and sustains psychological safety among its employees? A one-word answer to this vital and provocative question is simply "leadership." As suggested earlier in this chapter, the leaders of an organization promote, enable, or disenable the frank and open dialogue that reflects a particular level of psychological safety.

Let's consider six basic leadership qualities that have been empirically verified to distinguish the leaders of "good" versus "great" companies, and note their connection to encouraging and nurturing psychological safety. Many readers will recognize this reference to the classic best seller: "From Good to Great" by Jim Collins (2001).

More than two decades ago, Jim Collins and his research team studied 11 companies that had transitioned from being good to being great, and they had sustained this level of excellence for at least 15 years. The researchers systematically compared these good-to-great companies with a carefully selected set of 11 companies that maintained good productivity and profits for at least 15 years but never made the leap from good to great.

Collins and his research team found a number of common qualities among the good-to-great companies. Among those distinctions were the following leadership characteristics observed at the eleven "good-to-great" organizations. Please consider how each of these six leadership qualities relate to cultivating a work culture that facilitates psychological safety.

1. **Manifest personal humility or compelling modesty.** This leadership quality receives priority attention in Collins' book. He relates several case studies of companies that did not reach their potential because their leaders were more concerned about their own notoriety than the reputation of their company. The "good-to-great leaders never wanted to become larger-than-life heroes;" rather they "were seemingly ordinary people quietly producing extraordinary results" (p. 28).
2. **Project success beyond self.** Related to the first leadership quality, the most effective leaders attribute company success to factors beyond themselves. As systems thinkers, they see the big picture and realize their success is contingent on the daily small-win accomplishments of several individuals. They routinely acknowledge the synergistic contributions of the many employees who enable remarkable results.
3. **Accept responsibility for failure.** While spreading success beyond themselves, great leaders take full responsibility for organizational failures. They face the brutal facts of less-than-desired outcomes, and they hold themselves accountable without blaming other people or just "bad luck." Interestingly, Collins and his team found that the leaders

of the comparison companies often blamed others for lackluster performance while taking personal credit for extraordinary results. Social psychologists call this the "self-serving bias" (Mezulis et al., 2004).

4. **Promote a learning culture.** Humble leaders are open to new information. They are always learning, with a fervent belief in never-ending improvement. The good-to-great leaders observed by Collins' team facilitated fact finding to learn and improve, not to find someone to blame. They led with questions rather than answers, and they promoted frank and open dialogue and debate. The result: Employees were not satisfied with the status quo but were continuously engaged in finding ways to improve company performance. They were constantly alert to possibilities for process refinement—diligently searching for the best solutions to problems, and regularly submitting suggestions for fine-tuning their operations.
5. **Work to achieve, not to avoid failure.** Although they reveal and face brutal facts, the most effective leaders never waver in their resolve for greatness. Failure is not an option; it is not even considered. With an optimistic stance, these good-to-great leaders focused on achieving exemplary success. They attended to their envisioned enterprise with fanatic consistency and a disciplined constancy of purpose.
6. **Encourage self-motivation for meaningful work.** Self-motivation is key to safe and long-term productivity and this person-state is enhanced whenever perceptions of personal choice or autonomy, competence, and community or relatedness are heightened (Deci & Flaste, 1995; Deci & Ryan, 1995; Geller, 2016). Moreover, people are self-motivated when their behaviors provide natural ongoing consequences that are rewarding.

When does behavior on the job become intrinsically rewarding and self-motivating? Answer: When people believe their work is meaningful. When does this happen? Answer: Sometimes the special value of the effort is obvious, as when people are engaged in activities that prevent unintentional injuries. However, even in these cases, it is critical to provide interpersonal appreciation, gratitude, and supportive feedback to reassure people that they are accomplishing meaningful work. Great leaders know how to promote and support self-motivation, and they do that often.

This sixth leadership quality is especially significant for OSH, because it defines a self-transcendent source of self-directed motivation (Maslow, 1971). Specifically, working to keep people injury free is obviously meaningful AC4P behavior that fuels self-motivation. In this regard, Jim Collins ends his best-seller with the following: "It is impossible to have a great life unless it is a meaningful life. And it is very difficult to have a meaningful life without meaningful work" (p. 210). Safety leaders do meaningful work and therefore have meaningful lives.

19.4.3 Great Leaders and Psychological Safety

The obvious connections between the qualities of good-to-great leaders, defined more than two decades ago by Jim Collins and his research team, and the leadership qualities purportedly needed to inspire and support psychological safety are quite remarkable. Indeed, Susan Vargas (2021) concluded her introduction to psychological safety by listing five ways for anyone to cultivate psychological safety: 1) ask questions, 2) be humble, 3) build relationships, 4) value all perspectives, and 5) foster autonomy.

While managers and supervisors are assigned their corporate positions, any employee can choose to be a leader by practicing the six qualities of great leaders reviewed here as key to nurturing psychological safety. Please note also that these leadership qualities are relevant beyond the workplace to educational facilities, community-service agencies, sports teams, and in the home.

19.5 IN SUMMARY

This chapter began with a list of guidelines for initiating and sustaining a culture-improvement process aimed at achieving a TSC. Three support processes were identified to maintain employees' long-term commitment and involvement in a culture-improvement effort—leadership, communication, and recognition. Communication and recognition were discussed earlier in Chapter 13, so leadership received more attention here.

Psychological science has demonstrated that the best leaders are enthusiastic, honest, motivated, confident, analytical, informed, and flexible. Although it is common to see those characteristics described as permanent personality traits, it is certainly reasonable to assume they can be enhanced through education, communication, recognition, and involvement in a successful OSH process. Thus, while it is useful to look for "natural" leaders when selecting members of a Safety Steering Team, it is important to realize that leadership qualities could be suppressed in some people by their lack of empowerment or low sense of belonging. New safety processes and an eventual culture change might disclose leaders you did not know existed in the workforce.

Involvement is key to so many aspects of building a TSC, and it can be increased in many ways. Expect to see five levels of involvement:

1. Total engagement from innovators who see change as an opportunity to improve.
2. Individuals committed but not totally engaged until direction and support are given.
3. People, usually the majority, are ready to get involved but remain on the sidelines until prodded and encouraged by others.
4. Doubters who see change as a problem and use learned helplessness and cynicism as excuses to remain detached.
5. The active resisters who see change as an opportunity to resist, complain, and promote mistrust.

Active and passive resisters (Categories 4 and 5) should be ignored, if possible. Recognize and support those willing to try the new process. Employees totally engaged in the process (Category 1) need to help individuals who are committed but not yet totally immersed (Category 2). Then, these two groups can work with the majority (Category 3) who need examples to follow. You can see why it is important to cultivate leadership, communication, and recognition skills among the "true believers" in innovation. Turn those leaders loose, and they will be your best recruiters to build the base of support for a TSC.

Psychological safety was introduced as a human dynamic that is receiving considerable attention these days and is critical for cultivating an actively caring for safety culture and achieving a TSC. While this culture-related concept might seem new to some readers, this interpersonal dynamic actually appeared in the management literature at least 30 years ago. Plus, while the safety and productivity advantages of cultivating a work, educational, or home culture that facilitates and supports psychological safety are obvious and demand serious consideration, such a mission and vision are certainly not new. As explained herein, the need to cultivate a culture of interpersonal trust, interdependent collaboration, and continuous improvement has been a primary mission of OSH since the 1980s or perhaps earlier. Indeed, the leadership strategies proposed to enhance and support psychological safety are the same as those identified empirically more than 20 years ago as qualities practiced by the leaders of especially successful organizations.

20 Reviewing the Principles

This text summarized principles for understanding the human dynamics of OSH. When you use these principles to design, execute, evaluate, and continuously improve interventions to improve safety-related behaviors and attitudes, you are well on your way to achieving a Total Safety Culture.

If you want to get a good idea, get a lot of ideas.

(Linus Pauling)

"How can we translate these concepts into real-world application?"
"Would you please put your theory into procedures or practices we could follow in our plant?"

I have heard questions like those at each of the Deming workshops I attended. They seemed to disappoint Dr. Deming (1991, 1992), who would assert that the purpose of the seminar was to teach theory and principles, not specific procedures. It was up to the participants to return to their own organizations and devise specific methods and procedures that fit their culture. Dr. Deming stressed the need to start with theory and then customize relevant practices.

As Dr. Deming knew well, and as I have discussed throughout this text, lasting improvement is founded on evidence-based procedures that fit the culture of an organization. Outside consultants can be invaluable for teaching appropriate principles and facilitating an implementation process, but if most of the employees do not understand or believe in the principles, well-intentioned efforts will never take root.

20.1 FIFTY KEY PRINCIPLES

It all starts with evidence-based theory. This final chapter specifies 50 important principles that summarize the psychological science of OSH and lay the groundwork for cultivating and achieving a TSC. Yes, 50 sounds like a long list, but these principles will be familiar to you, since all of these were derived from information covered in this text.

I hope you will find this list useful as a review and as a starting point for developing your OSH-enhancement process. Some of the principles focus on design and implementation, while others explain why we often fail in OSH. Most of these can be used as guidelines for checking potential long-term benefits of a specific safety-improvement procedure. All of these will help you appreciate the complex human dynamics of safety and health promotion.

This is not a priority list. Do not read anything into the order of the principles. However, I sincerely hope you will teach these evidence-based principles to others and apply several to actively care for safety and eventually achieve a TSC.

20.1.1 Principle 1: Safety Should Be Internally—Not Externally—Driven.

It is common to hear employees talk about safety "for" OSHA. It often seems employers and employees "do" safety more to satisfy the mandates of this outside regulatory agency than for themselves. This translates into perceptions of top-down control and performing to avoid failure rather than to achieve success.

Ownership, commitment, and proactive behaviors are more likely when people work toward achieving their own goals, not the government's policies. As discussed in Chapter 19, how we define programs and activities can influence attitudes that shape involvement. It makes most sense to talk about OSH as a mission that is owned and achieved by the very people it benefits.

20.1.2 Principle 2: Culture Change Requires People to Understand the Principles and Know How to Use Them.

Chapter 9 distinguished between education and training and emphasized that long-term culture change requires both. Education focuses on theory or principles. Training gets into the specifics of how to turn principles into effective action. Role playing with supportive and corrective feedback is essential for training, because participants get direct behavioral instruction on how correctly they are executing a particular procedure and how they can improve their behavior.

20.1.3 Principle 3: Champions of a TSC Will Emanate from Those Who Teach the Principles and Procedures.

When people teach, they "walk the talk" and become champions of continuous improvement. After more than 45 years of safety consulting, it is clear to me that success depends on the presence of such leaders. I have seen no better way to develop champions of a campaign than to first teach relevant theory and method, then show how others can be instructors, and finally provide opportunities for colleagues and coworkers to teach each other.

20.1.4 Principle 4: Leadership Can Be Developed by Teaching and Demonstrating the Qualities of Effective Leaders.

Just because you believe in something does not guarantee you will be an effective champion of the cause. Leaders have certain characteristics, as discussed in Chapter 19, that can be taught and cultivated among others. People need to understand the principles behind effective leadership and the behaviors that reflect quality leadership. You can also learn by observing the leadership skills of others. When you see leaders in action, reward their exemplary behavior with quality recognition, sincere gratitude, and/or supportive feedback.

20.1.5 Principle 5: Focus Recognition, Education, and Training on People Reluctant but Willing, Rather Than on Those Actively Resisting.

As discussed in Chapter 19, people resist change for many reasons. Some feel insecure leaving their comfort zones. Some mistrust any change in policy or practice that was not their idea. Others appreciate the special attention they get from coworkers by resisting. It is usually a waste of time trying to force change on these folks. In fact, resistance often hardens as more pressure to change is applied.

20.1.6 Principle 6: Giving People Opportunities for Choice Can Increase Commitment, Ownership, and Engagement.

A basic reason for preferring the use of positive over negative consequences to motivate behavior (Chapter 11) is that people feel more free. They perceive more choice when working to achieve rewards than when working to avoid penalties. As illustrated in numerous laboratory experiments and field applications, increasing perceptions of choice leads to more self-motivation and engagement in the process (see Chapter 6).

20.1.7 PRINCIPLE 7: A TSC REQUIRES CONTINUOUS ATTENTION TO FACTORS IN THREE DOMAINS: ENVIRONMENT, BEHAVIOR, AND PERSON.

Early on, the "Safety Triad" was introduced, with behavior and person-states representing the psychology of OSH. That is the focus of this book, but do not overlook a need for environmental change. The environment includes physical conditions and the general atmosphere or ambiance regarding OSH. The behavioral safety analysis presented in Chapter 9 started with addressing ways to simplify the task through re-engineering. Thus, before addressing behavior change, it is critical to improve environmental conditions that can make a job more user-friendly and ergonomically sound.

20.1.8 PRINCIPLE 8: DO NOT COUNT ON COMMONSENSE FOR SAFETY IMPROVEMENT.

Common sense is not evidence based but is biased by an individual's subjective interpretation of personal experience. As a researcher of applied psychological science for more than 50 years, I have become quite committed to this basic principle (discussed in Chapter 2). Indeed, I have dedicated most of my career to discovering principles of human behavior through systematic application of the scientific method.

20.1.9 PRINCIPLE 9: SAFETY INCENTIVE/REWARD PROGRAMS SHOULD FOCUS ON THE PROCESS RATHER THAN ON OUTCOMES.

One of the most frequent commonsense mistakes in OSH is the use of outcome-based incentive/reward programs. Giving rewards for avoiding an injury seems reasonable and logical, but it readily leads to covering up minor injuries and a distorted perspective of safety-related performance. The basic activator–behavior–consequence contingency (see Chapter 8 and Principle 18) demonstrates that safety incentives need to focus on process activities, or safety-related behaviors. However, as illustrated in Figure 20.1, it is easy to become overly focused on outcome measures and overlook the processes needed to achieve the desirable outcome.

FIGURE 20.1 Focus on process to improve outcome statistics.

20.1.10 PRINCIPLE 10: SAFETY SHOULD NOT BE CONSIDERED A PRIORITY BUT A VALUE WITH NO COMPROMISE.

As discussed in Chapter 3, this is the ultimate vision—safety becomes a value linked to every priority in the workplace or wherever we find ourselves. Priorities change according to circumstances; values are deep-seated personal beliefs beyond compromise.

20.1.11 PRINCIPLE 11: SAFETY IS A CONTINUOUS FIGHT WITH HUMAN NATURE.

I have known people who meet the behavioral criteria for holding safety as a value—they practice safety, teach safety, go out of their way to actively care for the safety of others. Unfortunately, relatively few individuals meet these criteria. Why? Because human nature (natural or intrinsic motivating consequences, Chapter 11) typically encourages at-risk behavior.

People seek comfort, convenience, and expediency. When you compete with natural supportive consequences in order to teach, motivate, or improve behavior, you are fighting human nature. It takes emotional intelligence or EQ (Chapter 15) to overcome the natural tendency to work for soon, certain, and positive consequences, which often includes at-risk behavior.

20.1.12 PRINCIPLE 12: BEHAVIOR IS LEARNED FROM THREE BASIC PROCESSES: CLASSICAL CONDITIONING, OPERANT CONDITIONING, AND OBSERVATIONAL LEARNING.

Through naturally occurring consequences and planned instructional activities, we learn every day and we develop attitudes and emotional reactions to people, events, and environmental stimuli. The mechanisms for learning behavior and emotions voluntarily and involuntarily were explicated in Chapter 8.

20.1.13 PRINCIPLE 13: PEOPLE VIEW BEHAVIOR AS CORRECT AND APPROPRIATE TO THE DEGREE THEY SEE OTHERS DOING IT.

Because personal experience often convinces us that "it's not going to happen to me," we need a powerful reason to perform safely when personal injury is improbable. So, please consider this: Everyone who sees you acting safely or at risk could learn a new safe or at-risk behavior or perceive support for safe or at-risk behavior already learned. Now, consider the vast number of people who observe your behavior every day. Our influence as a social model gives us special responsibility to go out of our way for OSH.

20.1.14 PRINCIPLE 14: PEOPLE WILL BLINDLY FOLLOW AUTHORITY, EVEN WHEN THE MANDATE RUNS COUNTER TO GOOD JUDGMENT AND SOCIAL RESPONSIBILITY.

The fact that people often follow top-down rules without regard to potential risk is alarming. This puts special responsibility on managers and supervisors who give daily direction. Those front-line leaders could signal, even subtly, the approval of at-risk behavior in order to reach production demands. People are apt to follow even implicit demands from their supervisor to whom they readily delegate responsibility for any injury that could result from their at-risk behavior.

20.1.15 PRINCIPLE 15: GROUP PARTICIPATION CAN BE ENHANCED BY INCREASING PERSONAL RESPONSIBILITY, INDIVIDUAL ACCOUNTABILITY, GROUP COHESION, AND INTERDEPENDENCE.

Giving up personal responsibility for safety to another person (Principle 14) could be due to the lack of those factors that are needed for interdependent teamwork. Thus, workplace interventions

Reviewing the Principles

and action plans need to be implemented with the aim of increasing an individual's perception of individual accountability and personal responsibility, including one's sense of group cohesion and interdependence (Chapter 17).

20.1.16 PRINCIPLE 16: ON-THE-JOB BEHAVIORAL OBSERVATION AND INTERPERSONAL BEHAVIORAL FEEDBACK ARE KEY TO ACHIEVING A TSC.

Critical behavior checklists (Chapter 8) and communicating the results of checklist observations (Chapter 12) put this principle to work. Unlike the situation depicted in Figure 20.2, the observation and feedback process must be positive. Only then will this basic improvement tool spread throughout a work culture. The more people giving and receiving interpersonal behavioral feedback related to OSH, the greater the improvement in safety-related behavior and the more injuries prevented.

20.1.17 PRINCIPLE 17: BEHAVIOR-BASED SAFETY IS A CONTINUOUS DO IT PROCESS WITH D = DEFINE TARGET BEHAVIORS, O = OBSERVE TARGET BEHAVIORS, I = INTERVENE TO IMPROVE BEHAVIORS, AND T = TEST IMPACT OF THE INTERVENTION.

The four-step DO IT process enables continuous improvement through an objective behavior-focused approach. As detailed in Chapter 8, people need to decide on critical target behaviors to observe. After baseline observations are taken, a behavior-improvement intervention is developed and implemented. By continuing to observe the target behavior(s), the impact of the intervention process can be objectively evaluated. Results might suggest a need to refine the intervention, implement another intervention, or define another set of behaviors to target. The next four principles provide guidance for designing a behavior-improvement intervention.

FIGURE 20.2 While some negative feedback seems warranted, it is unlikely to have a beneficial effect.

20.1.18 Principle 18: Behavior Is Directed by Activators and Motivated by Consequences.

External or internal events occurring before behavior (referred to as activators) only motivate to the extent they signal or specify consequences. Intentions and goals can motivate behavior if they stipulate positive or negative behavioral consequences. Understanding this principle is critical for the development of effective behavior-improvement interventions.

20.1.19 Principle 19: Intervention Impact is Influenced by the Amount of Response Information, Participation, and Social Support, as Well as External Consequences.

Interventions that give specific instructions (response information) and get participants actively involved are likely to influence desirable behavior and a supportive attitude. If the intervention facilitates support from others, such as coworkers or family members, it can have lasting effects. Furthermore, people are more likely to develop internal motivation when external rewards or threats are relatively small and insufficient to completely justify the target behavior.

20.1.20 Principle 20: Extra and External Consequences Should Not Over-Justify the Target Behavior.

The various examples of positive consequences presented in Chapters 11 and 12 are neither large nor expensive. For the reasons discussed previously, rewards should not provide complete justification for desired behavior. We do not want people complying with safety rules only to gain a reward or to avoid a penalty. If that is the case, what happens when the consequences—good or bad—are taken away? The reason to comply is removed. This is why many employees use PPE at work, but rarely at home.

20.1.21 Principle 21: People Are Motivated to Maximize Positive Consequences (Rewards) and Minimize Negative Consequences (Costs).

Of course, this principle relates to many behaviors. In Chapter 14, this principle was used to explain why people often do not rush to help in a crisis. If there are more perceived costs than benefits to intervening, AC4P behavior is unlikely. Therefore, a prime strategy for increasing the frequency of safety-related behavior is to overcome the costs (negative consequences) with benefits (positive consequences). Various kinds of consequences are defined by the next principle.

20.1.22 Principle 22: Behavior Is Motivated by Six Categories of Consequences: Positive vs. Negative, Natural vs. Extra, and Internal vs. External.

Understanding these characteristics (as explained in Chapter 11) can enable significant insight into the motivation behind observed behavior. Appreciating those various consequences can also suggest whether an external intervention is needed to improve behavior and what kind of intervention should be implemented.

Can you define the type of consequences motivating the biker in Figure 20.3? The odometer provides external and natural immediate feedback to the exerciser as he pedals. When he talks to himself while pedaling, he adds internal motivating feedback to the situation. His evaluation of the feedback determines whether the feedback is positive or negative.

Reviewing the Principles 237

FIGURE 20.3 Natural immediate consequences can be very motivating.

20.1.23 Principle 23: Negative Consequences Have Four Undesirable Side-Effects: Escape, Aggression, Apathy, and Countercontrol.

How did you feel the last time you received a reprimand from a supervisor? Maybe, you felt like slinking away or taking a swipe at him or her. Chances are you did not go back to the job charged up with positivity. Perhaps you wanted to do something to make him or her look bad. These and other undesirable side-effects of using negative consequences were discussed in Chapter 11.

20.1.24 Principle 24: Natural Variation in Behavior Can Lead to a Belief That Negative Consequences Have More Impact Than Positive Consequences.

Behavior fluctuates from good to bad for many reasons. Seldom can peak performance be sustained, and poor performance is almost bound to get better at some point. So, if you praise someone and their performance falters, don't swear off positive supportive feedback, and don't overestimate the power of your reprimand if it seems to get some immediate results. The improvement could have been due to natural or "common cause" regression to the mean. Keep things in perspective. Remember, only with positive consequences can both behavior and attitude improve simultaneously.

20.1.25 Principle 25: Long-Term Behavior Change Requires People to Change "Inside" as Well as "Outside".

The psychological science of safety requires us to consider both external behavior and internal person-states. Chapter 15 focused on the role of person-states in influencing people to actively care for another person's safety and health. Chapter 16 showed how outside factors can be manipulated to influence those person-states and, thus, increase occurrences of AC4P behavior. A TSC reflects the integration and synergy of behavior-based and person-based psychology. The next several principles focus on understanding "inside" person-based factors.

20.1.26 PRINCIPLE 26: ALL PERCEPTION IS BIASED AND REFLECTS PERSONAL HISTORY, PREJUDICES, MOTIVES, AND EXPECTATIONS.

Appreciating this principle is key to understanding people and realizing the importance of empathic listening to others before intervening. It also supports the need to depend on objective, systematic observation for knowledge rather than commonsense (Principle 8).

It is important to realize the reciprocal relationship between perception and behavior. Perceptions influence actions and, in turn, actions influence perceptions. If we perceive a risk, we will act to reduce it; by acting to reduce a risk, we will become more aware of other risks.

20.1.27 PRINCIPLE 27: PERCEIVED RISK IS LOWERED WHEN A HAZARD IS PERCEIVED AS FAMILIAR, UNDERSTOOD, CONTROLLABLE, AND PREVENTABLE.

When people perceive a new risk, they adjust their behavior to avoid it. Call it "fear of the unknown." The reverse is also true. As discussed in Chapter 5, research has shown that hazards perceived as familiar, understood, controllable, and preventable are viewed as less risky. This is why many hazards are underestimated by employees.

20.1.28 PRINCIPLE 28: THE SLOGAN "ALL INJURIES ARE PREVENTABLE" IS FALSE AND REDUCES PERCEIVED RISK.

Frankly, I believe telling people all injuries are preventable insults their intelligence. They know better. It is difficult enough to anticipate and control all environmental and behavioral factors contributing to injuries, but controlling factors inside people is nearly impossible. The most critical problem with this popular slogan is that it can reduce the perception of risk. Hazards considered controllable and preventable are perceived as relatively risk free.

20.1.29 PRINCIPLE 29: PEOPLE COMPENSATE FOR INCREASES IN PERCEIVED SAFETY BY TAKING MORE RISKS.

As reviewed in Chapter 5, researchers have shown that some people will compensate for a decrease in perceived risk by performing more risky behavior. In other words, some people increase their tolerance for risk when feeling protected with a safety device. As shown in Figure 20.4, high technology safety engineering can give a false sense of security. This is not the case for people who hold safety as a value (Principle 10).

20.1.30 PRINCIPLE 30: WHEN PEOPLE EVALUATE AT-RISK BEHAVIOR OF OTHERS, THEY FOCUS ON INTERNAL FACTORS; WHEN EVALUATING THEIR OWN AT-RISK BEHAVIOR, THEY FOCUS ON EXTERNAL FACTORS.

As discussed in Chapter 6, this principle is termed "the fundamental attribution error." It contributes to systematic bias whenever we attempt to evaluate others, from completing performance appraisals to conducting an injury analysis. Because we are quick to attribute internal (person-based) factors to other people's behavior, we tend to presume consistency in others because of permanent traits or personality characteristics. To explain injuries to other persons, we use

Reviewing the Principles

FIGURE 20.4 Technology can cause reduced perception of risk and at-risk behavior.

expressions like, "He's just careless," "She had the wrong attitude," and "They were not thinking like a team."

However, when evaluating our own undesirable behavior, we point the finger to external factors. Figure 20.5 illustrates this bias in a context many readers can relate to from personal experience. This should make us stop and realize the many external variables that can be observed and often changed to increase everyone's safety-related behavior and reduce injuries throughout a culture.

FIGURE 20.5 It feels better to project our imperfections on outside factors.

20.1.31 Principle 31: When Succeeding, People Over-Attribute Internal Factors, but When Failing, People Over-Attribute External Factors.

This research-based principle is referred to as the "self-serving bias" (see Chapter 6) and is sure to distort an injury analyses, formerly called an "investigation" (Chapter 9). Placing blame for a mistake on outside variables is just a basic defense to protect one's self-esteem. In most organizations, even a minor injury is perceived as a failure. As a result, the victim is sure to avoid discussing inside, person factors contributing to a mishap. However, attributing our successes to internal personal factors enhances our self-esteem.

20.1.32 Principle 32: People Feel More Personal Control and Perceived Choice When Working to Achieve Success Than When Working to Avoid Failure.

The sense of having control over life events is one of the most important person-states contributing to subjective wellbeing (SWB). When we feel in control, we are more motivated and we work harder to succeed. We are also more likely to accept failure as something we can change. Thus, the value of increasing people's sense of personal control and perceived choice for safety-related interventions is obvious.

20.1.33 Principle 33: Stressors Lead to Positive Stress or Negative Distress Depending on Appraisal of Personal Control.

When we believe we can do things to reduce our stressors—work demands, interpersonal conflict, boredom—we are more motivated to take control. As discussed in Chapter 6, this is positive stress, an internal person-state not nearly as detrimental to OSH as distress. We feel distress when we believe there is little we can do about current stressors. This person-state can lead to frustration, exhaustion, burnout, and at-risk behavior.

20.1.34 Principle 34: In a Total Safety Culture Everyone Goes Beyond the Call of Duty for OSH—They Actively Care for Safety.

Here, we have a primary theme of this book. While applied behavioral science provides methods and techniques to improve the human dynamics of OSH, principles from person-based psychology (e.g., humanism) need to be considered to ensure the behavior-based tools are used. The ultimate aim is to integrate behavior-based and person-based psychology (e.g., humanistic behaviorism) so everyone participates in efforts to achieve a TSC. In the ideal culture, everyone actively cares for the safety and health of themselves and others.

20.1.35 Principle 35: Actively Caring Should Be Planned and Purposeful and Focus on Environment, Person, or Behavior.

Part 5 of this text reflects the title of this text—*Actively Caring for Safety*—from understanding why people resist actively-caring-for-people (AC4P) behavior (Chapter 12) to implementing strategies for increasing occurrences of AC4P behavior (Chapter 14). We need to plan ways to enable and nurture as much AC4P behavior as possible rather than sitting back and waiting for "random acts of kindness."

20.1.36 Principle 36: Direct, Behavior-Focused AC4P is Proactive and Most Challenging and Requires Effective Communication Skills.

Some acts of caring are relatively painless and effortless—contributing to a charity, sending a get-well card, or actively listening with empathy to another person's problems. Telling someone how to change his or her behavior can be confrontational and challenging, especially when it is direct. This is the type of AC4P behavior we are most likely to avoid, which is unfortunate because it is usually the most beneficial.

20.1.37 PRINCIPLE 37: SAFETY COACHING THAT STARTS WITH CARING AND INVOLVES OBSERVING, ANALYZING, AND COMMUNICATING LEADS TO HELPING.

The basic components of effective safety coaching were presented in Chapter 12, with each letter of COACH signifying a label for the sequence of events in the process. The coaching process should start with an atmosphere of interpersonal Caring and an agreement that the coach may Observe an individual's performance, preferably with a behavioral checklist. Then, the coach Analyzes the observations from a fact-finding, system-level perspective. Subsequently, the results are Communicated in a one-to-one AC4P conversation, with the sole purpose to Help another individual reduce the possibility of personal injury.

20.1.38 PRINCIPLE 38: ACTIVELY CARING CAN BE INCREASED INDIRECTLY WITH PROCEDURES THAT ENHANCE SELF-ESTEEM, BELONGING, AND EMPOWERMENT.

This principle reflects one of the most innovative and important theories presented in this book (Chapter 15). Substantial research is available to support each component of this principle. Procedures that enhance a person's sense of self-esteem ("I am valuable"), belonging ("I belong to a team"), and empowerment ("I can make a difference") make it more likely that a person will actively care for the safety and/or health of another person. Nourishing each of these person-states leads to the AC4P belief that "We can make valuable differences."

20.1.39 PRINCIPLE 39: EMPOWERMENT IS FACILITATED WITH INCREASES IN SELF-EFFICACY, RESPONSE-EFFICACY, AND OUTCOME EXPECTANCY.

When people's sense of self-efficacy ("I can do it"), response-efficacy ("It will work"), and outcome expectancy ("It is worth it") are increased, they are more likely to feel empowered ("I can make a difference") and perform relevant AC4P behaviors. Empowerment does not necessarily result from receiving more authority or responsibility.

In order to truly feel empowered, people need to perceive they have the skills, resources, and opportunity to take on the additional responsibility (self-efficacy), believe the intervention process they have learned to implement will enable the achievement of their goal (response-efficacy), and believe that the application of the intervention process is worth the time and effort required for implementation (outcome expectancy).

These three beliefs needed to feel empowered reflect *training* (knowing what to do), *education* (becoming aware of relevant research principles and/or empirical results that demonstrate effectiveness of the intervention), and *motivation* (anticipating the potential beneficial consequences that will follow a successful intervention process).

20.1.40 PRINCIPLE 40: WHEN PEOPLE FEEL EMPOWERED, THEIR SAFE BEHAVIOR SPREADS TO OTHER SITUATIONS AND BEHAVIORS.

In a TSC, people go beyond the call of duty for safety. This means they perform safe behavior in various situations. They show both stimulus generalization—performing a particular safe behavior in various settings—and response generalization—performing several OSH-related behaviors in a particular situation or setting. Figure 20.6 reflects both stimulus and response generalization of AC4P.

FIGURE 20.6 The best interventions spread their effects to other behaviors and environmental settings.

20.1.41 PRINCIPLE 41: AC4P BEHAVIOR CAN BE INCREASED DIRECTLY BY EDUCATING PEOPLE ABOUT FACTORS CONTRIBUTING TO BYSTANDER APATHY.

Chapter 16 explained strategies for encouraging AC4P behavior directly. This principle reflects a basic procedure for doing this. Educating people about the barriers to helping others can remove some obstacles to AC4P behavior and increase occurrences of AC4P behavior. Similarly, discussing the barriers to safe behavior can inspire some people to improve OSH, provided they have learned effective techniques for doing this.

20.1.42 PRINCIPLE 42: AS THE NUMBER OF OBSERVERS OF A CRISIS INCREASES, THE PROBABILITY OF HELPING DECREASES.

This principle is probably the first barrier to AC4P behavior that should be taught. It is strange but true and means that people cannot assume someone else will intervene in a crisis. In fact, the most common excuse for not actively caring for safety is something like, "I thought someone else would do it" or "I didn't know it was my responsibility." This principle reflects the need to promote a social norm that it is everyone's responsibility to actively care for safety.

20.1.43 PRINCIPLE 43: AC4P BEHAVIOR IS FACILITATED WHEN APPRECIATED AND INHIBITED WHEN UNAPPRECIATED.

Making an effort to actively care directly for someone else's safety is a big step for many people and deserves genuine gratitude and supportive feedback. Then, if advice is called for to make the AC4P behavior more effective, corrective feedback should be given appropriately. Be sure to make your emotional deposits first. All AC4P behavior is well intentioned, but it is not frequently accompanied by the kind of behavioral feedback that shapes improvement. A negative reaction to an act of caring can severely discourage the benefactor from trying again. Consequently, much of the future of AC4P behavior is in the hands of the relatively few beneficiaries who receive a benefactor's attempt to actively care.

20.1.44 Principle 44: A Positive Reaction to AC4P Behavior Can Increase the Benefactor's Self-Esteem, Empowerment, and Sense of Belonging.

This is a follow-up to Principle 43 and supports the need to sincerely recognize occurrences of AC4P behavior. Although research in this domain is lacking, it is intuitive that feeling successful at performing AC4P behavior should lead to more AC4P behavior. Success should enhance the benefactor's self-esteem, empowerment, and belonging and thereby increase the probability of more AC4P behavior. Thus, we have the potential to initiate a mutually supportive cycle of AC4P behavior, provided the beneficiary's reaction to AC4P behavior is positive.

20.1.45 Principle 45: The Universal Norms of Consistency and Reciprocity Motivate Everyday Behaviors, Including AC4P Behavior.

These two social influence norms have a powerful impact on human behavior. Sometimes, people apply these norms intentionally to influence others. At other times, these norms are activated without our awareness. Regardless of intention or awareness, behavior-change techniques derived from these norms can be very effective, as elucidated by the next three principles.

20.1.46 Principle 46: Once People Make a Commitment, They Encounter Internal and External Pressures to Think and Act Consistently with Their Position.

This is why you can act people into thinking differently or think people into acting differently. If people act in a certain way on the "outside," they will adjust their "inside"—including perceptions, beliefs, and attitudes—to be consistent with their behavior. The reverse is also true, but throughout this text I have recommended targeting behavior first because behavior is easier to change on a large scale than is thinking.

Figure 20.7 depicts a humorous scenario of rational behavior preceding the internal emotion of fear. Does this seem realistic? Is it reasonable to believe that the act of running from a bear

FIGURE 20.7 Behavior precedes emotion; we are afraid because we run.

will come before an internal person-state? In fact, this is likely what happens, as predicted by the James-Lange theory of emotion. As James-Lange (1890) put it, "We feel sorry because we cry, angry because we strike, and afraid because we tremble" (p. 1066).

20.1.47 Principle 47: The Consistency Norm Is Responsible for the Impact of the "Foot-in-the-Door" Technique.

As detailed in Chapter 16, the "foot-in-the-door" technique of social influence succeeds because of the consistency norm (Principle 46). When an individual agrees with your relatively small request (e.g., to serve on a safety committee), you have your foot in the door. To be consistent, that person is more likely to agree later with your larger request, perhaps to give a safety presentation at a plant-wide meeting. Similarly, when people sign a petition or promise card that commits them to act in a certain way—perhaps to actively care for the safety of others—they experience pressure from the consistency norm to follow through.

20.1.48 Principle 48: Managers Hold People Accountable; Leaders Inspire Self-Accountability.

Management is not the same as leadership, but both are invaluable for OSH. Managers occupy assigned positions and hold employees accountable to perform certain behaviors. In contrast, leaders emanate naturally from all levels of an organization and inspire people to be self-accountable or self-motivated. Everyone can learn how to practice effective leadership for OSH, including managers.

Systematic research has revealed that the perceptions of personal choice (Chapter 6), competence, and community or interdependence (Chapter 16) enhance an individual's self-accountability or self-motivation. People can inspire these perceptions within themselves and among others. Individuals who influence the perception of choice, competence, and/or community among others are the most effective leaders. My 15-minute TEDX talk on self-motivation explains how to make that happen. That presentation was posted in December 2013 and had 12 million views in January 2024.

Consider showing that TEDX talk to a group of individuals and afterwards ask them to operationalize the leadership principles presented by defining situations and personal interactions in their culture that can promote and/or support perceptions of choice, competence, and community among participants. Just Google my name and TEDX to review this popular teaching/learning tool.

20.1.49 Principle 49: When Numbers from Program Evaluations are Meaningful to the Participants, They Can Direct and Motivate Intervention Improvement.

This principle relates to the critical issue of program evaluation (Chapter 18). In safety, the total recordable injury rate is the most popular evaluation number used to rank companies for safety rewards. It is calculated by multiplying the number of workplace injuries by 200,000 and then dividing the answer by the total person-hours worked in that time period. This is an obvious example of an abstract number with little meaning. The most direct measure of ongoing safety performance comes from behavioral observations, and Chapters 8, 12, and 18 recommended ways to obtain meaningful feedback numbers from process evaluations.

If your objective is to increase risk awareness and motivate safe behavior, the most influential evaluation tool you can use is actually anecdotal. The most moving feedback usually comes from the personal report of an injured employee. Therefore, the work culture needs to support the reporting of close calls and personal injuries, as well as the discussion of strategies for preventing future undesirable incidents. For this to happen throughout an organizational culture, "psychological safety" is needed—our last principle.

20.1.50 PRINCIPLE 50: A TOTAL SAFETY CULTURE REQUIRES PSYCHOLOGICAL SAFETY.

In psychologically safe organizations, participants feel accepted and respected and believe they will not be humiliated or embarrassed when speaking up with questions, concerns, mistakes, or new ideas. Employees who feel psychologically safe are more engaged in their work; they speak up more often and are motivated to improve the performance of their work team and the entire organization. The managers and leaders of an organization set the cultural tone of that organization, thereby enabling or disenabling interpersonal trust and the frank and open dialogue that reflects a particular level of psychological safety.

Six leadership qualities that promote and support psychological safety were illustrated, as determined by systematic empirical investigation, and each quality is intuitive and consistent with commonsense. Specifically, anyone, including a manager, can be a leader and cultivate psychological safety by: 1) manifesting personal humility or compelling modesty, 2) projecting group success beyond oneself, 3) accepting responsibility for failure, 4) promoting a learning culture, 5) working to achieve success, not to avoid failure, and 6) encouraging self-motivation (Principle 48) for meaningful work.

20.2 IN SUMMARY

This chapter reviewed the principles of human dynamics discussed throughout this text. Founded on research published in scientific journals, they enable profound understanding of the psychological science of OSH and AC4P behavior. Use these as guidelines to develop, implement, evaluate, and refine safety-improvement strategies and you will make a positive difference in the OSH of your organization, community, or culture.

Champions are needed to lead this process. Some are easy to find; others will evolve when the principles reviewed here are taught. Give potential champions opportunities to teach these principles to others and to help develop intervention procedures for promoting and supporting OSH. Active participation increases both the belief in the principles and the empowerment to apply them to achieve a TSC.

There is no quick fix to achieve a TSC. A continuous OSH-improvement journey will include some bumpy roads, forced detours, and missed turns. These 50 principles are your map to reach an enviable destination, but be prepared to blaze new paths and traverse difficult terrain. Please do not forget to take a break now and then to appreciate the achievement of journey milestones. Recognize with sincere gratitude those behaviors that contribute to a successful journey.

At the end of the second Deming workshop I attended, a participant raised his hand to ask one final question. When acknowledged, he stood and walked to the nearest microphone and stated: "Dr. Deming, you have taught us many important principles to consider when designing procedures to transform a culture. But frankly, this challenge seems overwhelming. Can we really expect to make a difference in our lifetime?"

W. Edwards Deming, at age 92, replied, "That's all you've got!"

Glossary of Key Terms

ABC (Activator–Behavior–Consequence): the three-term contingency of applied behavioral science (ABS), specifying that a stimulus event (an activator) occurring before a behavior provides direction (e.g., an incentive or a disincentive), whereas the consequence following a behavior provides motivation to repeat or stop the preceding behavior (e.g., a reward or penalty).

Activator: an environmental event (e.g., directions from a sign or a person, an incentive or a disincentive) implemented to influence the occurrence of a particular target behavior.

Actively Caring: the combination of action and compassion, resulting in behavior performed on behalf of the safety, health, and/or wellbeing of one or more other persons.

Actively Caring for People (AC4P): the application of behavioral science and select principles from humanism to increase the frequency and/or improve the quality of behavior that benefits human health, safety, security, and/or welfare—referred to as humanistic behaviorism.

AC4P Coaching: interpersonal communication whereby one individual (the teacher, coach, or parent) employs principles of humanistic behaviorism to benefit the observed behavior of another individual (the learner).

Applied Behavioral Science (ABS): the application of research-based principles derived from experimental behavior analysis to increase the occurrence of desirable behaviors and to decrease the frequency of undesirable behaviors.

Attributional Bias: a prejudiced assumption of why a person is acting in a certain way—reactive thinking as opposed to reflective thinking.

Authority Principle: the tendency to comply with a request from a person with presumed credibility and perhaps influence over relevant behavioral consequences.

Behavior-Based Feedback: interpersonal communication following the occurrence of an observed behavior that shows the performer what s/he did correctly (supportive feedback) and/or incorrectly (corrective feedback).

Behavior-Based Safety (BBS): the systematic application of applied behavioral science to prevent personal injury by increasing the occurrence of safe behavior and decreasing the frequency of at-risk (or unsafe) behavior, typically manifested by a peer-to-peer behavioral-observation-and-feedback process.

Belongingness: perceived mutual social support or interdependency with others—friends, family, colleagues, or even strangers.

Bystander Apathy Effect: the tendency for any given observer of a person needing help to be less likely to give aid if other bystanders are present, presumably due to diffusion of personal responsibility.

Classical Conditioning: a learning process whereby a previously neutral stimulus comes to elicit (or cause) an automatic or reactive response (or behavior)—an involuntary conditioned response.

Cognitive Dissonance: tension or discomfort experienced when an individual perceives his/her behavior to be inconsistent with his/her attitude, belief, and/or values, or vice versa.

Community: the perception of relatedness or belongingness with others—often an experience of interdependent social or group support regarding the accomplishment of a certain challenge or assignment.

Compassion: AC4P action or behavior following empathy—a sincere concern for another person's welfare or wellbeing.

Glossary of Key Terms

Conditioned Stimulus: a stimulus previously neutral with regard to triggering a particular response acquires the capacity to elicit that same response (a conditioned response) through the learning process of classical conditioning.

Consequence: an environmental event (positive or negative) that naturally occurs (i.e., intrinsic) after a behavior, or is added to the situation (i.e., extrinsic) in order to influence a certain behavior, which may or may not influence the frequency or improve the quality of the preceding behavior.

Contingency Management: manipulating a preceding activator and the consequence(s) of a target behavior in order to increase (with an incentive/reward) or decrease (with a disincentive/penalty) its quality or its frequency of occurrence.

Corrective Feedback: behavior-focused information given to an individual after s/he performs an undesirable or less-than-optimal behavior, often manifested in a one-to-one coaching interaction.

Countercontrol: the occurrence of a behavior that is contrary to a desired target behavior because the person perceives a loss of personal choice or individual freedom as the result of a top-down rule, a disincentive, or an if-then contingency.

Critical Behavior Checklist (CBC): a list of safe and at-risk behaviors defined by a work team and used to systematically record observations of safe and at-risk behaviors during a peer-to-peer behavioral-observation-and-feedback session.

Disincentive: an activator (or extrinsic stimulus) that announces the delivery of a certain negative consequence (a penalty) following the occurrence of a designated unwanted behavior.

Disposition: an internal trait or person-state of an individual that predisposes or influences his/her behavior in a particular situation.

Distress: the physiological and psychological reaction to an event that is perceived as important—a stressor—but appraised as threatening or overwhelming and beyond the person's current domain of personal control.

DO IT Process: an acronym reflecting the basic scientific method of applied behavioral science: "D" for Define the target behavior to influence, "O" for Observe the natural occurrence frequency of the target behavior (i.e., baseline), "I" for Intervene to change the frequency of the target behavior in a desirable direction (i.e., increase or decrease), and "T" for Test the impact of the intervention by observing frequencies of the target behavior before, during, and after the intervention ends in order to evaluate the impact of the intervention.

Education: teaching one or more individuals the theory, principles, and procedures related to a certain task or the circumstances related to that task.

Emotion: a potent and potentially long-term reaction to a person or a situation that can motivate relevant self-directed behavior.

Emotional Intelligence (EQ): one's ability to: 1) give up a soon, certain, and positive consequence for a delayed and uncertain but larger positive consequence (i.e., impulse control); 2) remain in control and optimistic following personal failure and frustration (i.e., intrapersonal EQ); and 3) understand and empathize with other people and work with them cooperatively (i.e., interpersonal EQ).

Empathic Listening: the highest level of interpersonal listening whereby the listener attempts to identify with and understand the speaker's feelings, attitude, motives and behavior, as well as the circumstances influencing relevant dispositions and behavior.

Empathy: attempting to identify with and understand another individual's circumstances that influence his/her feelings, attitude, motives, and behavior.

Empowerment: an individual's belief or confidence that s/he can master a particular assignment or accomplish a certain task, determined by answering "Yes" to three questions—Can

I do it? (self-efficacy), Will it work? (response efficacy), and Is it worth it? (outcome expectancy).

Extrinsic: a stimulus event external to an individual, considered an activator or a consequence when applied to influence (or motivate) the occurrence of a target behavior.

Extrinsic Motivator: an activator (e.g., an incentive or disincentive) that influences the occurrence of a behavior in order to gain a positive consequence, avoid a negative consequence, or escape an aversive environmental event.

Extrinsic Reward: A positive behavioral consequence delivered after the occurrence or outcome of behavior(s) in order to support the behavior(s) and/or to enhance an individual's sense of competence or self-efficacy.

Extrinsic Stimulus: an observable and external environmental event that might direct behavior (as an activator) or motivate behavior (as a positive or negative consequence).

Failure Avoider: an individual performing a certain behavior in order to avoid a negative consequence.

Feedback: behavior-focused information (supportive or corrective) following the performance of a behavior, delivered to influence the frequency and/or the quality of the preceding behavior.

Fundamental Attribution Error: the tendency for an observer of another person's behavior to underestimate the impact of the context (situational factors) and to overestimate the impact of dispositional factors (person-states and/or personality traits).

Gratitude: an expression or a reception of sincere appreciation for an individual's behavior that temporarily boosts the positive person-state of subjective wellbeing for both the benefactor and the beneficiary.

Humanism: an appreciation of empathy, diversity, and human dignity; and a sincere concern for human wellbeing.

Humanistic Behaviorism: making applications of applied behavioral science implemented to improve human performance more effective by practicing empathy, and considering such person-states as self-esteem, self-efficacy, belongingness, personal control, and optimism.

If-Then Reward: a positive behavioral consequence whereby an activator had announced the delivery of a positive consequence following the occurrence of a certain desired behavior—an if-then reward contingency.

Incentive: an announcement (an activator) that specifies the availability of a positive consequence (a reward) following the performance of a designated behavior or a desirable outcome of more than one behavior.

Interdependence: a collectivistic, we-need-each-other viewpoint or mindset that facilitates and supports AC4P behavior—acts performed on behalf of the health, safety, security, and/or wellbeing of others.

Intervention: an external or extrinsic program or process implemented to influence the quantity or quality of one or more target behaviors.

Intrinsic Consequence: a positive or negative stimulus event that flows naturally after a behavior and keeps it going (a positive reinforcer) or causes it to decrease in frequency or to stop completely (a punisher).

Leadership: inspiring people to be self-motivated and to feel empowered to perform one or more goal-directed behaviors (i.e., discretionary behavior).

Management: motivating one or more behaviors to occur through an external or extrinsic accountability system (e.g., activator, behavioral feedback, incentive, disincentive, rule, mandate, interpersonal recognition or reprimand).

Negative Reinforcement: a procedure or situation in which the frequency or form of a behavior is increased in order to avoid or escape an undesirable extrinsic stimulus.

Glossary of Key Terms

Nondirective: practicing empathic listening to learn another person's perceptions and relevant circumstances before providing advice or direction.

Now-That Reward: a positive behavioral consequence (e.g., behavior-based recognition or supportive feedback) delivered after observing the occurrence of a desirable behavior—given to express sincere appreciation for the behavior and to potentially influence the recipient's perceived competence and self-motivation, as well as to increase the probability of the desired behavior recurring.

Observational Learning: initiating or improving the performance of one or more behaviors by watching one or more individuals perform the behavior(s).

Occupational Safety and Health (OSH): the target domain for the evidence-based intervention principles and procedures explained and illustrated in this text, as derived from psychological science.

Optimism: the expectation that the consequences of one or more behaviors will be positive—a challenge will be handled successfully.

Other-Directed Behavior: behavior that occurs as a result of an extrinsic or external intervention, such as an incentive that promises a reward following a target behavior or a disincentive that warns the levy of a penalty if a certain undesirable behavior occurs.

Outcome Expectancy: anticipating the positive consequence(s) to follow the occurrence of one or more goal-directed behaviors and believing that achieving the consequence is worth the effort and time involved—the response cost.

Penalty: a negative consequence following a designated behavior or an outcome of one or more behaviors that was intended to decrease recurrence of the behavior(s).

Perceived Choice: the perception of having more than one option with regard to accomplishing a particular task or action plan.

Percent-Safe Score: the percentage of safe behaviors observed by systematically completing a critical behavior checklist during a peer-to-peer behavioral observation-and-feedback session, calculated by dividing the total number of safe behaviors observed by the total number of behavioral observations (safe plus at-risk behaviors) and multiplying the quotient by 100.

Performance: the output or outcome of a process or system which includes input from behavior, dispositions (person-states and personality traits), and situational factors.

Personal Control: one's perception of having the ability or competence to accomplish a particular task successfully—feeling empowered to meet a particular challenge.

Person-State: a personal disposition (e.g., expectancy, attitude, or personality characteristic) that influences an individual's behavior, but varies as a function of the situation or environmental context.

Positive Reinforcement: a procedure or situation that increases the frequency or improves the form of a behavior by following that behavior with a desirable consequence—termed a positive reinforcer.

Premature Cognitive Commitment: the notion that people maintain a preconceived perception about someone or something (i.e., inflexible prejudice) as the result of prior personal experience, attitudes, cognitions, and/or behaviors.

Principle of Reciprocity: the social norm that obligates people to repay others with similar behavior (positive or negative) they have received from them—"pay it forward" or "pay back."

Psychological Reactance: the perception of a loss of personal choice or freedom due to a top-down rule, disincentive, or if-then contingency that activates the performance of behavior contrary to the desired behavior targeted by the rule, disincentive, or if-then contingency.

Psychological Safety: a critical human dynamic whereby an individual feels accepted and included in a particular organizational culture and believes s/he will be respected when speaking up with questions, concerns, mistakes, or new ideas.

Punisher: a negative consequence following a designated behavior that reduces the occurrence of that behavior.

Punishment: a behavior-focused intervention that successfully reduces the frequency of a target behavior by following that behavior with a negative consequence.

Reinforcer: an extrinsic (extra) or intrinsic (natural) stimulus or environmental event that increases the frequency and/or the form of the behavior it follows.

Response Efficacy: an individual's belief or confidence that performing one or more goal-directed behaviors will contribute to achieving a desired outcome, mission, or long-term vision.

Reward: a positive consequence following a designated behavior or an outcome of more than one behavior that may or may not influence recurrence of the behavior(s).

Selection by Consequences: one of B.F. Skinner's legacies that reflects the motivational principle that people do what they do in order to gain a positive consequence or to avoid or escape a negative consequence.

Selective Perception: the process of interpreting and organizing sensory information that is biased by prior experience and/or expectations and may influence attitude, thinking, and behavior.

Self-Accountability: internal motivation (from within an individual) to perform one or more self-directed behaviors.

Self-Actualization: the top of Maslow's *initial* Hierarchy of Needs at which a person feels a sense of ultimate achievement—having fulfilled one's potential.

Self-Directed Behavior: self-motivated behavior that is not solely motivated by an external or extrinsic motivator (e.g., an incentive or a disincentive).

Self-Efficacy: an individual's belief or confidence that s/he has the knowledge, skill, and ability to perform a goal-directed behavior needed to accomplish a certain task.

Self-Esteem: a general or overall feeling of self-worth that influences one's propensity to perform AC4P behavior.

Self-Fulfilling Prophecy: a prediction or expectation that directly or indirectly becomes true because relevant behavior is influenced in order to confirm the prophecy.

Self-Motivation: a person-state that reflects internal drive to perform a certain behavior or achieve a particular outcome of one or more behaviors from a self-directed and self-accountability perspective.

Self-Serving Bias: taking personal credit for our successes but blaming our failures on situational factors beyond our personal control in order to maintain a positive view of ourselves and protect our self-esteem.

Self-Talk: intrapersonal conversation (or covert verbal behavior) that can influence overt behavior, and often reflects and/or affects one's attitude, perception, self-esteem, and/or self-motivation.

Self-Transcendence: the top of Maslow's *revised* Hierarchy of Needs whereby an AC4P mindset is realized, and the individual experiences personal fulfillment and positive reinforcement when performing AC4P behavior—behavior contributing to the health, safety, security, or wellbeing of one or more persons.

SMARTS Goal: a behavior-focused achievement objective defined by making it **S**pecific, **M**otivational, **A**ttainable, **R**elevant, **T**rackable, and **S**hared.

Stress: the physiological and psychological reaction to an event perceived as important and challenging—a stressor—and appraised as within one's current domain of personal control and empowerment.

Subjective Well-Being (SWB): a positive person-state of perceived happiness or life satisfaction influenced by interpersonal and intrapersonal conversation, among other contributing factors.

Success Seeker: an individual performing one or more behaviors in order to earn a positive consequence, which can be intrinsic or extrinsic.

Supportive Feedback: a now-that reward (or positive recognition) delivered after observing an individual's desirable behavior, which is most rewarding and influential when it can SOAR—it is **S**pecific (designates the behavior); **O**n time (occurs soon after the behavior is observed); **A**ppropriate for the knowledge, ability, and experience of the performer; and shows **R**eal appreciation for the individual's competence and effort.

Synergy: a process of interdependent participation in a goal-directed activity that results in greater achievement than possible from all participants performing alone and independently.

Total Safety Culture (TSC): an environmental context or setting (e.g., a workplace, school, or home) where people interact daily on behalf of the health, safety, security, and wellbeing of everyone else in the surroundings with a spirit of win/win interdependence and self-transcendence.

Training: providing behavior-focused direction and feedback regarding the quality of one or more specified behaviors performed to achieve a designated outcome.

Trait: a disposition to feel and act in a particular way, presumed to be a stable life-long personality characteristic resulting from nature rather than nurture.

Type A Personality: a disposition (or person-trait) reflecting people who are competitive, ambitious, hard-driving, impatient, aggressive, and anger prone.

Type B Personality: a disposition (or person-trait) reflecting people who are easygoing, relaxed, and laid back—capable of living in the present and enjoying the moment.

Vicarious Punishment: observational learning whereby the negative consequence given to one or more persons following undesirable behavior influences an observer of this interaction to decrease the occurrence of that behavior.

Vicarious Reinforcement: observational learning whereby the positive consequence given to one or more persons following a desirable behavior influences an observer of this interaction to increase the frequency or improve the quality of his/her performance of that behavior.

Vision: a distant or ultimate objective or aspiration, presumed to be reached or achieved through ongoing SMARTS goal-setting and the successive achievement of the relevant SMARTS goals.

References

American Heritage Dictionary. (1985). (2nd College ed.). Boston, MA: Houghton Mifflin company.
Bandura, A. (1982). Self-efficacy mechanisms in human agency. *American Psychologist, 37*, 122–147.
Bandura, A. (1997). *Self-efficacy: The exercise of control*. New York: W. H. Freeman.
Batson, C. D., Bolen, M. H., Cross, J. A., & Neuringer-Benefiel, H. E. (1986). Where is altruism in the altruistic personality? *Journal of Personality and Social Psychology, 1*, 212–220.
Bird, F. E., Jr., & Germain, G. L. (1997). *The property damage accident: The neglected part of safety*. Loganville, GA: Institute Publishing.
Brown, H. J., Jr. (1991). *Life's little instruction book*. Nashville, TN: Rutledge Hill Press.
Bugelski, B. R., & Alimpay, D. A. (1961). The role of frequency in developing perceptual sets. *Canadian Journal of Psychology, 5*, 205–211.
Carnegie, D. (1936). *How to win friends and influence people* (1981 ed.). New York: Simon & Schuster, Inc.
Centers for Disease Control and Prevention. (2023, May 4). *Current cigarette smoking among adults in the United States*. Atlanta, Georgia: Centers for Disease Control.
Cialdini, R. B. (2001). *Influence: Science and practice* (4th ed.). New York: Harper Collins College.
Clark, T. R. (2020a). *The 4 stages of psychological safety: Defining the path to inclusion and innovation*. San Francisco: Berrett-Koehler Publishers.
Clark, T. R. (2020b, October 13). To foster innovation, cultivate a culture of intellectual bravery. *Harvard Business Review*. https://hbr.org/2020/10.
Collins, J. (2001). *From good to great*. New York: Harper Collins.
Covey, S. R. (1989). *The seven habits of highly effective people: Powerful lessons in personal change*. New York: Simon & Schuster.
Daniels, A. C. (2000). *Bringing out the best in people* (2nd ed.). New York: McGraw-Hill.
Deci, E. L., & Flaste, R. (1995). *Why we do what we do: Understanding self-motivation*. New York: Penguin Books.
Deci, E. L., & Ryan, R. M. (1995). *Intrinsic motivation and self-determinism in human behavior*. New York: Plenum.
Dembroski, T. M., & Costa, P. T., Jr. (1987). Coronary prone behavior: Components of the Type A pattern and hostility. *Journal of Personality, 55*, 211–235.
Deming, W. E. (1985). Transformation of western style of management. *Interfaces, 15* (3), 6.
Deming, W. E. (1986a, July). *Drastic changes for western management*. Abstract for the meeting of TIMS/ORS at Gold Coast City, Australia.
Deming, W. E. (1986b). *Out of the crisis*. Cambridge, MA: Massachusetts Institute of Technology, Center for Advanced Engineering Study.
Deming, W. E. (1991). *Quality, productivity, and competitive position*. Four-day workshop presented in Cincinnati, Ohio, by Quality Enhancement Seminars, Inc.
Deming, W. E. (1992). *Quality concepts to solve societal crises: Profound knowledge for psychologists*. Washington, DC: Invited Keynote Address at the Centennial Convention of the American Psychological Convention.
Deming, W. E. (1993). *The new economics for industry, government, education*. Cambridge, MA: Massachusetts Institute of Technology, Center for Advanced Engineering Study.
Dinwiddie, F. W. (1975). Humanistic behaviorism: A model for rapprochement in residential treatment milieus. *Child Psychiatry and Human Development, 5* (4), 254–259.
Drebinger, J. W., Jr. (2000). *Mastering safety communication: Communication skills for a safe, productive and profitable workplace* (2nd ed.). Galt, CA: Wulamoc Publishing.
Editors of Conari Press. (1993). *Random acts of kindness*. Emeryville, CA: Conari Press.
Festinger, L. (1957). *A theory of cognitive dissonance*. Evanston, IL: Row & Peterson.
Frankl, V. (1962). *Man's search for meaning: An introduction to logotherapy*. Boston: Beacon Press.
Fulghum, R. (1988). *All I really need to know I learned in kindergarten*. New York: Ivy Brooks.
Gardner, H. (1993). *Multiple intelligences*. New York: Basic Books.
Geller, E. S. (1994). Ten principles for achieving a total safety culture. *Professional Safety, 39* (9), 18.
Geller, E. S. (2000). A new method for motivating employees. *Industrial Safety & Hygiene News, 34* (10), 20, 22–23.
Geller, E. S. (2008). The tragic shooting at Virginia Tech: Personal perspectives, prospects, and preventive potentials. *Traumatology, 14* (1), 8–20.

References

Geller, E. S. (2016). The psychology of self-motivation. In E. S. Geller (Ed.). *Applied psychology: Actively caring for people* (pp. 83–118). New York: Cambridge University Press.

Geller, E. S. (2020). Humanistic behaviorism. In A. Sharman (Ed.). *One percent safer: The secrets to achieving excellence from the world's finest thinkers* (pp. 82–83). South Wales, Great Britain: Maverick Eagle Press.

Geller, E. S. (2022). Psychological safety: The optimal context for injury prevention. *Professional Safety, 67* (1), 18–20.

Geller, E. S. (2023). Emotional intelligence: A crucial human dynamic for occupational safety and health. *Professional Safety, Professional Safety, 28* (12), 24–39.

Geller, E. S. (2023). If you see something, say something: Cultivating an actively-caring-for-people culture. *Behavior and Social Issues, 32*(8), 482–485.

Geller, E. S., Bruff, C. D., & Nimmer, J. G. (1985). The "flash for life": A community prompting strategy for safety belt promotion. *Journal of Applied Behavior Analysis, 18*, 145–159.

Geller, E. S., & Geller, K. S. (2023). *The human dynamics of achieving an injury-free workplace: Safety directives from psychological science* (2nd ed.). Newport, VA: GellerAC4P, Inc.

Geller, E. S., Perdue, S., & French, A. (2004). Behavior-based safety coaching: Ten guidelines for successful application. *Professional Safety, 49* (7), 42–49.

Goleman, D. (1995). *Emotional intelligence.* New York: Bantam Books.

Goleman, D. (1998). *Working with emotional intelligence.* New York: Bantam Books.

Grote, D. (1995). *Discipline without punishment.* New York: American Management Association.

Guastello, S. J. (1993). Do we really know how well our occupational accident prevention programs work? *Safety Science, 16*, 445.

Haddon, W., Jr. (1968). The changing approach to the epidemiology, prevention, and amelioration of trauma: The transition to approaches etiologically rather than descriptively based. *American Journal of Public Health, 58*, 1431.

Hall, E. T. (1966). *The hidden dimension.* Garden City, NY: Doubleday.

Hansen, L. (1994). Rate your B.O.S.S.: Benchmarking organizational safety strategy. *Professional Safety, 39* (6), 37.

Hasanzadeh, S., de la Garza, J. M., & Geller, E. S. (2020). Latent effect of safety interventions. *Journal of Construction Engineering and Management, 146* (5), 1–13.

Heinrich, H. W. (1931). *Industrial accident prevention: A scientific approach.* New York: McGraw-Hill.

Heinrich, H. W., Petersen, D., & Roos, N. (1980). *Industrial accident prevention: A safety management approach* (5th ed.). New York: McGraw-Hill.

Houston, B. K., & Vavak, C. R. (1991). Cynical hostility: Developmental factors, psycho-social correlates, and health behaviors. *Health Psychology, 10*, 9–17.

James, W. J. (1890). *Principles of psychology.* New York: Holt.

Jones, J. W. (1984). Cost evaluation for stress management. *EAP Digest,* 34.

Kanfer, F. H., & Phillips, J. S. (1970). *Learning foundations of behavior therapy.* New York: Wiley.

Kaufman, S. B. (2020). *Transcend: The new science of self-actualization.* New York: Penguin Random House, LLC.

Langer, E. J. (1989). *Mindfulness.* Reading, MA: Penguin Random House, LLC.

Langer, E. J. (1997). *The power of mindful learning.* Reading, MA: Perseus Books.

Latané, B., & Darley, J. M. (1970). *The unresponsible bystander: Why doesn't he help?* New York: Appleton-Century-Crofts.

Lloyd, S. R. (1996). *Leading teams: The skills for success.* West Des Moines, IA: American Media.

Ludwig, T. D., & Geller, E. S. (1997). Assigned versus participative goal-setting and response generalization: Managing injury control among professional pizza deliverers. *Journal of Applied Psychology, 82* (2), 253–261.

Mager, R. F., & Pipe, P. (1997). *Analyzing performance problems or you really oughta wanna* (3rd ed.). Atlanta, GA: The Center for Effective Performance.

Maslow, A. H. (1943). A theory of human motivation. *Psychological Review, 50*, 370–396.

Maslow, A. H. (1971). *The farther reaches of human nature.* New York: Viking.

Mezulis, A. H., Abramson, L. Y., Hyde, J. S., & Hankin, B. L. (2004). Is there a universal positivity bias in attributions? A meta-analytic review of individual, developmental, and cultural differences in the self-serving attributional bias. *Psychological Bulletin, 130* (5), 711–747.

Milgram, S. (1963). Behavioral studies of obedience. *Journal of Abnormal Social Psychology, 67*, 371.

Milgram, S. (1974). *Obedience to authority.* New York: Harper Collins.

New Merriam-Webster Dictionary. (2023). Springfield, MA: Merriam-Webster Inc., Publishers.

Norman, D. A. (1988). *The psychology of everyday things*. New York: Basic Books.

Oliver, M., Browning, S. L., Davis, M., & Geller, E. S. (2021). Face-mask wearing and social distancing: A test of risk compensation theory. *The Virginia Journal of Business, Technology, and Science, 1* (1). https://doi.org/10.51390/vajbts.v1i1.17

Parker, G. M. (Ed.). (1996). *The handbook of best practices for teams* (Vol. 1). Amherst, MA: HRD Press.

Pavlov, I. P. (1927). *Conditional reflexes*. (G. V. Anrep, Editor and Translator). London: Oxford University Press.

Peck, M. S. (1979). *The different drum: Community making and peace*. New York: Simon & Schuster.

Piliavin, J. A., Dovidio, J. F., Gaertner, S. L., & Clark, R. D. (1981). *Emergency intervention*. New York, NY: Academic Press.

Rees, F. (1997). *Teamwork from start to finish*. San Francisco, CA: Jossey-Bass.

Sandman, P. M. (1991). *Risk 5 hazard + outrage: A formula for effective risk communication*. A videotape presented to the American Industrial Hygiene Association, Environmental Communication Research Program, Cook College, Rutgers University, New Brunswick, NJ.

Seligman, M. E. P. (1975). *Helplessness: On depression development and death*. San Francisco, CA: Freeman.

Selye, H. (1974). *Stress without distress*. Philadelphia, PA: Lippincott.

Skinner, B. F. (1938). *The behavior of organisms*. Acton, MA: Copley Publishing Group.

Skinner, B. F. (1974). *About behaviorism*. New York: Alfred A. Knopf.

Skinner, B. F. (1981). Selection by consequences. *Science, 213*, 477–481.

Slovic, P. (1991). Beyond numbers: A broader perspective on risk perception and risk communication. In D. Mayo, & R. Hollander (Eds.). *Deceptable evidence: Science and values in risk management*. New York: Oxford.

Sulzer-Azaroff, B., & Austin, J. (2000). Does BBS work? Behavior-based safety and injury prevention: A survey of the evidence. *Professional Safety, 45* (7), 19–24.

Thoresen, C. E. (1972, April). *Behavioral humanism*. Research and Development Memorandum No. 88. Stanford, CA: School of Education, Stanford University.

U.S. Department of Labor. (2023). *Ergonomics—overview*. Washington, DC: Occupational Safety and Health Administration.

Vargas, S. (2021). Psychological safety: A hot concept, can it help create safer workforces? *Safety+Health, 203* (6), 42–45.

Watson, D. L., & Tharp, R. G. (1997). *Self-directed behavior: Self-modification for personal adjustment* (7th ed.). Pacific Grove, CA: Brooks/Cole.

Wilde, G. J. S. (1994). *Target risk*. Toronto, Ontario, Canada: PDE Publications.

Yerkes, R. M., & Dodson, J. D. (1908). The relation of strength of stimulus to rapidity of habit formation. *Journal of Comprehensive Neurological Psychology, 18*, 459–482.

Index

A

ABC, *see* Activator-Behavior-Consequence (ABC) model
AC4P coaching, 176
    person-based approach, *see* person-based approach to actively caring
    psychology of bystander apathy, 166–168
    social responsibility norm, 168
        role in Total Safety Culture, 218, 240, 245
    summary, 171–172
accident investigation, 29
accident proneness, 6, 22
accountability *vs.* responsibility, 103
Activator-Behavior-Consequence (ABC) model
    DO IT process and, 83
    human behavior and, 17, 99
    operant conditioning and, 77
    safety coaching process and, 145–146
activators, *see also* Activator-behavior-consequence (ABC) model
    behavior influenced by, 17, 77
    intervening with, *see* intervening with activators
actively caring
    behavior-based efforts example, 163
    belonging and, 178–179
    categorizing behaviors, 163
    consequences and, 161, 168–170, 189
    context considerations
        definition of context, 170–171
        safety at work and, 170–171
    continuous improvement and, 162
    described, 162
    education and training influences, 189
    factors involved, 161–162
    hierarchy of needs, 165–166
    increasing, *see* increasing actively caring
Actively Caring for People (AC4P) Movement, 161, 240, *see also* AC4P coaching
"Actively Caring Thank-You Cards," 189
active resisters, 224, 226
aggression and perceived control by negative consequences, 125
Airline Lifesaver intervention example, 113–114
Alimpay, D.A., 47
"all injuries are preventable" slogan, 38, 186, 238
All I Really Need to Know I Learned in Kindergarten (Fulghum), 119
apathy and perceived control by negative consequences, 125
applied behavioral science, 95
arousal and performance, 59
"As you know" phrase, 151
at-risk behavior, *see also* calculated risks
    feedback process, 90
    inadvertent rewards, 96–97
    motivational interventions and, 101–102
    reduction of, 64, 68
attitude, 37
attributional bias
    fundamental error of, 67
    self-serving bias, 68, 240, 250
authority principle, 40–43, 246
autobiographical bias, 151
automatic *vs.* controlled processing, 87–88
automobile safety
    safety-belt use
        Airline Lifesaver, 113–114
        cost effectiveness, 71
        Flash for Life, 112–113
avoidance contingency, 80

B

BASIC ID, 36–38, 43–44
Batten, Joe, 22
behavior
    factors in safety culture, 26–27
    factors in systems approach, 28–29
    perceived risk and, 48–49
    psychological dimension of, 36
behavioral safety analysis
    behavior-based feedback, 95, 98, 104, 133, 202
    behavior-based safety (BBS), 99–100
    behavior-based training, 99–100
    critical behavior identification, *see* critical behaviors identification
    intervention and behavior change accountability *vs.* responsibility, 103
        flow of behavior change, 102–103
        intervention strategies, 100–102
        types of behavior, 101
    percent safe score, 92
    reducing behavioral discrepancies consequences used effectively, 97
        inadvertent at-risk rewards, 96–97
        inadvertent punishment, 95–96
        in job matching, 98
        quick fix assessment, 95
        in skills, 97
        summary, 103–104
        task simplification, 93–94
        in training, 98
behavior-based approach/programs
    actively caring and, 161–162
    cost effectiveness of, 18–20
    described, 3
    human element considerations, 6–8
    learning from experience
        classical conditioning, 75
        definition of learning, 75–76
        observational learning, 78–79
        operant conditioning, 76–78
        overlapping of types, 79–80
    need for, 28–29
    person-based approach *vs.* behavior-based, 16
        cost effectiveness, 18–20

described, 16, 18
integration of, 19–20
primacy of behavior
at-risk behavior reduction, 72–73
increasing safe behavior, 74–75
training, 108
treatment or prevention, 71
response specificity impact, 114
validity of approach, 10
behavior change from interventions, *see also* safety coaching
accountability *vs.* responsibility, 103
approaches, 102
discipline use, 125–127
flow of behavior change, 102–103
intervention strategies, 101–102
types of behavior, 101
behavior observation in coaching critical behavior checklist
development, 137
feedback importance, 140–141
guidelines, 133
observation process features, 139–140
observation scheduling, 137–139
belongingness
actively caring behavior and, 179–181
emotional intelligence and, 179–180
group-based behavior and social loafing, 234
synergy, 187
bias
attributional, *see* perception
fundamental error of, 67
self-serving, 68, 240, 250
autobiographical, 151
contextual, 46–47
past experience and, 47–48
role in Total Safety Culture, 237
in supportive conversation, 151–152
Bird, Frank, 210
Blanchard, K., 8
Boydston, Steven, 114
Branson, R.M., 8
Brown, H.J., Jr., 163, 164
Bugelski, B.R., 47
bystander apathy
example, 165–166
research on, 167–168
role in Total Safety Culture, 240

C

calculated risks, *see also* at-risk behavior
emotional intelligence and, 179–180
human error and, 36
self-directed behavior and, 101
capture errors, 38–39
Carnegie, Andrew, 82
Carnegie, Dale, 119, 158
CBC, *see* critical behavior checklist
challenge, 61
checklist, *see* critical behavior checklist
choice, personal
risk perception and behavior and, 44
role in Total Safety Culture, 237
Cialdini, R.B., 190

Clark, R.D., III, 169
classical conditioning, 75
close call, 7, 28–30, 35–38, 43, 49, 50, 56, 67, 68, 72, 74, 85, 96, 97, 110, 117, 137, 200, 210, 211, 244
COACH acronym, 134
coaching, *see* safety coaching; supportive conversations
cognition, 37
cognitive dissonance, 65–66
cognitive failures and safety
"all injuries are preventable" slogan, 38, 186, 238
capture errors, 38–39
description errors, 39
loss-of-activation errors, 39
mistakes and calculated risks, 40
mode errors, 39
commitment and consistency
public and voluntary commitment, 190
role in Total Safety Culture, 240
in self-motivation personal/interpersonal pressures, 190
commonsense, 8
communication in coaching, *see also* safety coaching; supportive conversations
empathy, 143
to enhance others' self-esteem, 183–184
individual feedback, 143–144
interpersonal zone respect, 142
name use, 143
compassion, 246
compensation for risk
description and implications, 52–53
role in Total Safety Culture, 218
competency states, 102
complacency and risk, 50
comprehensive ergonomics
described, 4–5, 23
success of, 6–7
Conari Press, 161, 163, 164
conditioned response (CR), 80
conditioned stimulus (CS), 75, 80
conformity, *see* social conformity
consensus building for teamwork, 197–199
consequences, *see also* Activator-Behavior-Consequence (ABC) model
activator interventions and goal-setting, 116–118
for actively caring, 173–174, 180–183, 189
control by, 10
DO IT process and, 82–84
effectiveness of, 97
effects of, 18
emotional reactions to, 77–78
extrinsic, 120–122
incentives *vs.* disincentives, 115–116
interpersonal factors and, 96
intervention, using, *see* intervening with consequences
motivational interventions and, 101–102
role in Total Safety Culture, 237
selection by, 76–77
contests for safety, 129
context and actively caring
definition of context, 170
safety at work and, 170–171
contextual bias, 46
contingency management, 247

continuous improvement evaluation
    costs and benefits of safety, 214
    issues process development
        environmental conditions, 209–211
        person factors, 211–213
        process outcome cause-and-effect, 208
        process purpose, 204
        scope of measurements, 209
        work practices, 211
    role in Total Safety Culture, 237, 245
control
    by consequences, 10
    emotional intelligence and, 179–180
    empowerment and, 188
    role in Total Safety Culture, 218
    strategies for enhancing, 224
conversation, supportive, *see* supportive conversations
"The Cookie Thief," 135
corrective action plan, 126
corrective feedback
    openness to, 99
    in supportive conversation, 152
cost/benefits
    of behavior-based programs, 17–19
    benefits of safety, 228–229
    savings from safety-belt use, 111
countercontrol, 125, 237
couple at costume ball perception example, 44–45
Covey, Stephen R., 8, 48, 134, 143, 151, 189
Cox, Valerie, 135
CR (conditioned response), 80
critical behavior checklist (CBC)
    comments section importance, 93
    driving behaviors example, 89
    feedback process, 89
    in safety coaching development, 137
        feedback importance, 140
        observation process features, 139–140
        observation scheduling, 137–139
critical behaviors identification
    DO IT process
        basic approaches, 89–92
        behavior observation, 86
        consequences, 83–84
        described, 85
        intervention steps, 84
        role in Total Safety Culture, 237
        target behaviors, 84–85
    driving behaviors example
        activators and consequences, 87–88
        critical behavior checklist use, 88–89
        DO IT process and, 82–84
        multiple behaviors, 85–86
        skill maintenance training, 98
        target behaviors definition, 84–85
CS (conditioned stimulus), 75, 80
culture change, *see also* Total Safety Culture
    cultivating continuous support, 221–223
    education and training process set up course content, 219
    effectiveness evaluation, 220
    instructional process planning, 219–220
    instructor selection, 219
        evaluation procedures development, 218
    involvement importance, 225
    management support need, 218, 220

    overcoming resistance
        active resisters, 224
        level of participation, 225
        teaching skills needed, 224
        types of resistance, 225–226
    paradigm shifts for
        accident proneness mentality, 6, 22
        achievement orientation need, 25–26
        behavior focused approach, 26–27
        bottom-up involvement, 27
        continuous improvement, 30
        corporate responsibility and, 24–25
        fact-finding focus, 28–29
        interdependence, 28
        old three Es, 22–23
        paradigm definition, 24
        proactive, 30
        systems approach, 28
        value based, 30
    safety steering team creation, 218–218
    sustaining the process, 220

D

danger compensation, *see* risk, compensation
Daniels, Aubrey C., 130
Darley, John, 166, 167
Dawson, R., 9
defensive working style, 120
Deming, W. Edwards, 23, 28, 100, 107, 116, 117, 207, 215, 231, 245
Department of Transportation, U.S., 23
description errors, 39
The Different Drum: Community Making and Peace (Peck), 178
diffusion of responsibility and actively caring, 167, *see also* responsibility for safety
directive conversations, 158
discipline as intervention, 125–127
discrepancies, reducing behavioral
    consequences used effectively, 97
    inadvertent at-risk rewards, 96–97
    inadvertent punishment, 96–97
    quick fix assessment, 95
    in skills, 97
    task simplification, 93–94
    in training, 98
discriminative stimulus, 80
disincentive-penalty programs
    activator effectiveness and, 107
    behavioral-based safety and, 97
    disincentive, 97, 247
distress definition, 56
documentation of safety team meetings, 221
Dodson, J.D., 58, 59
DO IT process
    basic approaches, 89–92
    behavior observation and, 86
    consequences, 86–88
    described, 85
    interventions steps, 84
    role in Total Safety Culture, 237
    target behaviors, 84–85
Dovidio, J.F., *169*
Drebinger, J.W., Jr., 151

driving behaviors
    activators and consequences, 87–88
    critical behavior checklist use, 88–89
    skill maintenance training and, 98
drugs and safety, 37

E

education, *see also* training
    approach to safety, 23
    culture change process course content, 219
    effectiveness evaluation, 220
    instructional process planning, 219–220
    instructor selection, 219
    training *vs.*, 93
Eisenhower, Dwight, 147
Eliot, George, 173
Emerson, Ralph Waldo, 35
emotional bank account concept, 134
emotional intelligence
    impulse control and, 179–180
    intra- *vs.* interpersonal, 196–197
    relevance to occupational safety, 197
emotional reactions to consequences, 77–78
empathic listening, 143, 205, 238, 247
empathy, 142–143
empowerment
    approach to safety, 23–24
    described, 188
    optimism, 187
    personal control, 185–186
    role in Total Safety Culture, 241
    self-efficacy, 188
        response-efficacy, 176, 181, 223
enforcement approach to safety, 23–24
engineering changes approach to safety
    described, 5, 22
    task simplification from, 98
environmental factors
    actively caring and, 171
    evaluation conditions, 209–211
    safety culture role, 15–16
    systems approach and, 28
ergonomics approach to safety described, 4, 23
    success of, 6
errors from cognitive failures capture, 38–39
    description, 39
    judgment, 38
    loss-of-activation, 39
    mode, 39
escape and perceived control by negative consequences, 125
esteem, *see* self-esteem
evaluation approach to safety, 24, 218
execution factors and at-risk behavior reduction, 96–97
experience and bias, 46–47
expertise and leadership, 223
external consequences, 122
external locus of control, 189
extrinsic consequences, 120–122
Exxon Chemical, 130, 138

F

failure avoider, 248
fear-arousing approach to safety, 124

feedback
    behavior-based, 98
    communication in coaching and, 147
    corrective feedback
        openness to, 77
        in supportive conversation, 152
    from critical behavior checklist, 88, 137
    natural consequences, 120
    role in Total Safety Culture, 237
    skill maintenance training and, 98
    from tasks, 121
    on team performance, 202–203
Flash for Life example, 112–113
Flexibility and leadership, 223
Frankl, Viktor, 165, 238
frustration, 212
Fulghum, Robert, 119, 120
Fundamental Attribution Error, 238, 248

G

Gaertner, S.L., *169*
Gardner, H., 179
Geller, E.S., 53, 66, 101, 103, 113, 116, 176, 180, 182, 204, 205, 227, 229
Genovese, Kitty, 166–169
Germain, G.L., 210
goal setting, 9, 117
goal statements, 25
Goleman, D., 179, 180
government safety regulations, 24–25, 73
gratitude, 156
    expressing gratitude, 156
Grote, D., 126
group-based behavior
    belonging and
        social loafing, 234
        synergy, 188
    consensus building for teamwork, 197–199
    ineffectiveness of contingencies, 127
    problem solving, 5
    role in Total Safety Culture, 240
Guastello, Stephen, 4

H

habituation and habits, 101–103, 109–110
Haddon, William, 22, 23
Haddon, W., Jr., 22–23
Hall, E.T., 142
Hansen, Larry, 207
harm, 61
hazards and risk, 51
Heinrich, H.W., 72
Heinrich's Law, 72
HELP acronym in coaching, 144–146
Hierarchy of Needs, 165–166
Hoechst Celanese safety program, 130–131
honesty and integrity and leadership, 222
human element in safety, *see also* person factors in behavior
    barriers to safety, 50
    cognitive failures
        "all injuries are preventable" slogan, 38, 186, 238
        capture errors, 38–39

Index

description errors, 39
loss-of-activation errors, 39
mistakes and calculated risks, 40
mode errors, 39
considerations in behavior-based programs, 6–7
humanism, 17, 19, 20, *see also* person-based approach
humanistic behaviorism, 10–12, 18–21, 240, 248
human nature
    barriers to safety, 35–36
    dimensions of, 36–38
    perceived risk and safety, 49
    role in Total Safety Culture, 237
    in workplace injury, 4–6
hypocrisy effect, 66

I

if-then reward, 248
imagery, 37
impulse control, 180–181
incentive/reward programs
    activator effectiveness, 107
    in behavior-based approach, 16, 97
    material reward use considerations, 128
    motivational interventions and, 101–102
    outcome focused programs, 26–27
    reciprocity principle, 189–190
    role in Total Safety Culture, 237
    safety rewards considerations
        guidelines for effective approaches, 127–130
        program example, 130
    safety thank-you cards use, 130–131
    tangibles use, 154
    types of ineffective approaches, 127
incident analysis, 98–99, *see also* behavioral safety analysis
increasing actively caring
    direct effects on behaviors
        commitment and consistency, 190–192
        consequences and, 189
        education and training, 189
        reciprocity principle, 189–190
    person state enhancement strategies
        belonging, 187–188
        importance of feelings, 184
        optimism, 187
        personal control, 185–186
        self-efficacy, 184–185
        self-esteem, 183–184
injury prevention
    "all injuries are preventable" slogan, 38, 186, 238
    property damage incidents and, 200
instructional intervention, 101
instructor selection and culture change, 219, 230
intelligence, emotional, *see* emotional intelligence
interdependence and safety, 29
intermittent reinforcement schedule and motivation, 121
internal consequences, 122, 129
internal locus of control, 189
International Safety Rating System (ISRS), 5
interpersonal communication, 151, *see also* supportive conversations
interpersonal conversation, *see* supportive conversations
interpersonal factors in safety, *see also* person factors in behavior
    authority principle, 40–43, 246
    barriers of conformity and authority, 40
    consequences and, 95
    peer influence, 40–41
    psychological dimension of, 36
    trust and, 27
interpersonal intelligence, 179
interpersonal zone respect in coaching, 147
intervening with activators
    behavior specificity principle, 108
    consequence implication
        activator effectiveness and, 107
        goal setting, 117
        incentives *vs.* disincentives, 115–116
    habituation *vs.* salience, 109
    poster/sign use, 107–108, 110–111
    target audience involvement
        Airline Lifesaver, 113–114
        Flash for Life, 112–113
        ownership-involvement principle, 111–112
    timing and placement considerations, 114
intervening with consequences managing for safety
    behavior-consequence contingencies, 130
        discipline and involvement, 125–127
        negative consequence avoidance, 125
        power of consequences internal *vs.* external, 122
    intrinsic *vs.* extrinsic, 120–122
        simple rules importance, 119–120
    safety rewards considerations
        guidelines for effective approaches, 127–130
        program example, 130
    safety thank-you cards use, 130–131
    types of ineffective approaches, 127
interventions
    behavior change from
        accountability *vs.* responsibility, 103
        approaches, 102
        discipline use, 125–127
        flow of behavior change, 102–103
    intervention strategies, 101–102
    role in Total Safety Culture, 237
    supportive conversation use, *see* supportive conversations
    target audience involvement
        Airline Lifesaver, 113–114
        Flash for Life, 112–113
        ownership-involvement principle, 111–112
    timing and placement considerations, 114
    types of behavior, 101
    discipline used as, 125–127
intrapersonal communication
    described, 147
    verbal commitment indications, 149
intrapersonal intelligence, 179
intrinsic motivation, 120
intrinsic *vs.* extrinsic consequences, 120–122
ISRS (International Safety Rating System), 5

J

James-Lange Theory of Emotion, 244
James, William, 158, 244
job hazard analyses, 85, 120, 137

Johnson, D., xx
judgment errors, 38

K

Kennedy, John F., 133
knowledge factors and at-risk behavior reduction, 72

L

Langer, E.J., 54, 223
Lasorda, Tommy, 195
Latané, Bibb, 166, 167
leadership
    qualities of, 228–229
    role in Total Safety Culture, 43, 221–222, 232
learned helplessness, 63
learning
    from experience
        classical conditioning, 75–76
        definition of learning, 75
        observational learning, 78–79
        operant conditioning, 76–78
        overlapping of types, 79–80
        role in Total Safety Culture, 234
    from successes *vs.* failures, 152–153
Lewin, Kurt, 12
listening and coaching, 144–145
Lloyd, S.R., 205
locus of control, 189
loss-of-activation errors, 39–40
lottery incentive programs, 128
Lubbock, John, 44

M

Mager, R.F., 96
maintaining involvement in safety, 225
management
    actively caring and, 188
    matching talent to jobs, 98
Maslow, Abraham, 165–166, 229
Melville, Herman, 161
Michener, James, 217
Milgram, Stanley, 41–42
Mine Safety and Health Administration (MSHA), 24, 85
mission statement development, 12–13, 15, 20
mistakes and calculated risks, 40
mode errors, 39
monitoring achievement to increase safe behavior, 74
mood states, 168
motivation
    achievement orientation need, 25–26
    intermittent reinforcement schedule and, 121
    internal, 122, 132, 236, 250
    leadership and, 222
    requesting recognition, 156–157
    role in Total Safety Culture, 236, 244–245
    self-directed behavior
        commitment and consistency in, 190–192
        described, 101
        quality recognition and, 155–156
    self-reinforcement need, 121

motivational intervention, 101–102
motor vehicles, *see* automobile safety
MSHA (Mine Safety and Health Administration), 24, 85

N

name use in coaching, 143
National Highway Traffic Safety Administration (NHTSA), 22
natural consequences and at-risk behavior, 96–97
natural feedback, 120–121
    at-risk behavior reduction and, 72–73
    close call incidents approach to safety, 6
    personal blame removal, 211
    reporting, 6, 56
    target behaviors and, 84–85
negative attitudes, 73
negative consequences
    calculated risks and, 101
    compliance with activators and, 84
    reasons to avoid, 124–126
    role in Total Safety Culture, 236–237
negative reinforcement, 73, 248
    nondirective, 150–152, 223
    from tasks, 120–123
NHTSA (National Highway Traffic Safety Administration), 22
Nightingale, E., 9
Norman, Donald A., 38, 39
now-that rewards, 249, 251

O

obedience studies, 41–43
observational learning, 78–79
obtaining involvement in safety, *see* culture change
Occupational Safety and Health Administration (OSHA), 24, 26, 85, 116, 214, 231
one-to-one safety coaching, 90
operant conditioning, 76–78
optimism
    emotional intelligence and, 179–180
    empowerment and, 176–178
    resistance to distress, 63
    strategies for enhancing, 187
other-directed behavior, 101
outcome expectancy, 181, 223, 241
ownership-involvement principle, 111

P

Parker, G.M., 205
passion and leadership, 222
Pauling, Linus, 231
Pavlov, I.P., 75
Peck, M. Scott, 178
peer influence
    active resisters and, 226
    on safety-related behavior, 40–41
penalties, *see* punishment approach to safety
people factor, *see* person factors in behavior
perceived risk
    choice and, 50
    consequences acceptability and, 51–52

Index

familiarity and complacency, 50
hazard qualities and, 51
publicity and, 50
real risk *vs.*, 49
role in Total Safety Culture, 238
sense of fairness and, 52
sympathy for victims and, 50
percent safe behavior, 107, 140
percent safe score, 89, 92, 249
perception, *see also* bias
  contextual bias and, 46–47
  described, 44
  past experience bias and, 47–48
  of helplessness, 212
  perceived risk
    choice and, 50
    consequences acceptability and, 51–52
    familiarity and complacency, 50
    hazard qualities and, 51
    publicity and, 50
    real risk *vs.*, 49
    sense of fairness and, 52
    sympathy for victims and, 50
  premature cognitive commitment, 54, 151
  relevance to safety culture, 48
  risk compensation, 52–53
  self-efficacy and, 184
performance appraisals
  as stressors, 60
  for teamwork, 202–203
personal blame removal, 211
personal control
  emotional intelligence and, 179–180
  empowerment and, 176
  role in Total Safety Culture, 238
  strategies for enhancing, 185–186
personal protective equipment (PPE)
  CBC feedback process and, 92
  development of, 22
personal zone respect in coaching, 142
person-based approach/programs
  actively caring and, *see* person-based approach to actively caring
  behavior-based approach *vs.* cost effectiveness, 18–19
  challenges of, 18
  described, 16–17
  integration of, 19
person-based approach to actively caring
  behavior change basis, 173
  emotional intelligence
    impulse control and, 180–181
    intra- *vs.* interpersonal intelligence, 179–180
    relevance to occupational safety, 181
    measuring person-states, 212–213
    overview, 161, 173
  states of actively caring
    belonging, 178–179
    empowerment, 175–176
    self-esteem, 175
    traits *vs.* states, 174
person factors in behavior, *see also* human element in safety; interpersonal factors in safety
  cognitive failures
    "all injuries are preventable" slogan, 38, 186, 238
    capture errors, 38–39
    description errors, 39
    loss-of-activation errors, 39
    mistakes and calculated risks, 40
    mode errors, 39
  human nature and safety
    barriers to safety, 35–36
    dimensions of, 36–38
    role in Total Safety Culture, 234
  improvement evaluation process and, 211–213
  interpersonal authority's power, 41–43
    barriers of conformity and authority, 40–41
    peer influence, 40–41
    safety culture role, 15–16
    stressors coping mechanism, 61–67
    systems approach and, 28
personnel management, 95
person-states
  enhancement strategies
    belonging, 187–188
    importance of feelings, 182
    optimism, 187
    personal control, 185–186
    self-efficacy, 184–185
    self-esteem, 183–184
  improvement evaluation and, 212–213
  measuring, 212–213
  vs. traits, 174
perverse compensation, *see* risk, compensation
pessimist *vs.* optimist distinction, 63, 176–177
Petersen, D., 72
physical factors and at-risk behavior reduction, 72
physical fitness and stress, 63–64
Piliavin, J.A., 169
Pipe, P., 96
pop psychology
  humanism and, 17
  praise in public philosophy, 154
  self-help strategies shortcomings, 8–10
  self-motivation and, 119
positive consequences
  actively caring and, 189
  calculated risks and, 101
positive discipline as intervention, 126–127
poster/sign campaigns and safety
  impact of, 6
  as part of incentive program, 127
  as part of intervention, 107, 110–111
  safety-belt use impacted by, 114
  variation need, 110
  worker-designed slogan use, 132
PPE, *see* personal protective equipment
praise in private philosophy, 154
prejudgment filters, 150–151
premature cognitive commitment, 54, 151
primary appraisal of stressors, 60
proactivity, 30, 73
progressive discipline as intervention, 126
property damage incidents impact on workplace, 210
Proxemics, 169
psychological approach to safety, 22
psychological reactance, 23, 55
psychological safety, 199, 227–230, 245
psychotherapy approach, 18–19

publicity and risk, 50
punishment approach to safety
   inadvertent, 96
   reasons to avoid, 23, 124–126
   vs. reward approach, 76, 78

Q

questionnaires/surveys, 212, 213

R

radar detector use and habituation, 125
"random acts of kindness" concept, 161
reactive approach, 73
real risk vs. perceived risk, 49
reciprocity principle
   actively caring behavior and, 189–190
   recognition and, 156
   role in Total Safety Culture, 243
Rees, F., 205
renewing a team, 204
resistance to change, overcoming
   active resisters, 226
   level of participation, 225
   teaching skills needed, 223–224
   types of resistance, 225–226
response-efficacy, 176–176, 181, 223
response specificity impact, 108
responsibility for safety
   accountability vs., 103
   actively caring and, 167
   corporate responsibility and, 24–25
   diffusion of, 167
restructuring a team, 203–204
reward vs. punishment, 76, 78, see also incentive/reward programs
risk
   at-risk behavior
      feedback process, 92
      human error and, 40
      inadvertent rewards, 96
      motivational interventions and, 101
      reduction of, 72–73
   calculated
      emotional intelligence and, 180
      self-directed behavior and, 101
   compensation description and implications, 52–53
      role in Total Safety Culture, 238
   perceived
      choice and, 50
      consequences acceptability and, 51–52
      familiarity and complacency, 50
      hazard qualities and, 51
      publicity and, 50
      real risk vs., 49
      role in Total Safety Culture, 238
      sense of fairness and, 52
      sympathy for victims and, 50
risk homeostasis, see compensation for risk
"road rage," 57, 170, 178
Rogers, Will, 56
Roos, N., 72
Rotter, J.B., 209, 292

S

safe behavior opportunity (SBO), 91
safe behavior promise, 112, 190, 192
safety-belt use
   Airline Lifesaver, 113–114
   buckle-up road signs, 114
   cost effectiveness, 93
   Flash for Life, 112–113
safety belt use, safety coaching, see also supportive conversations
   action plan guidelines, 146
   communication in coaching to enhance others' self-esteem, 183–184
   eye contact use, 143
   individual feedback, 143–144
   interpersonal zone respect, 142
   name use, 143
      description of coaching, 135–136
   one-to-one with CBC, 90
   process of
      ABC analysis, 141
      behavioral observation, see behavior observation in coaching
      caring attitude, 134–136
   role in Total Safety Culture, 241
Safety Culture Survey, 212, 213
safety improvement programs
   approach selection
      human element, 6–8
      research evaluation, 3–6, 15
      research programs descriptions, 5–6
      self-help strategies shortcomings, 8
      seminar shortcomings, 7
   research importance
      behavioral science application, 10
      common beliefs refutation, 9
safety rewards, see incentive/reward programs
safety share
   activator effectiveness and, 110
   to increase safe behavior, 74–75
safety steering team
   creation for culture change, 218
   management support need, 220
Safety Thank-You Cards, 130–131
Safety Triad
   described, 71, 162
   evaluation process and, 209
   human nature and, 36
   role in Total Safety Culture, 233
salience and habituation, see habituation
Sandman, Peter M., 49
SBO (safe behavior opportunity), 91
secondary appraisal of stressors, 60
second hand recognition use, 154–155
selection by consequences, 76–78, 161
selective listening, 151
selective perception, 250
selective sensation, see perception
self-accountability, 103, 147, 204, 244
self-actualization, 165–166
self-confidence and leadership, 222–223
self-directed behavior
   commitment and consistency in, 190–192
   described, 101

Index

quality recognition and, 155–158
requesting recognition, 157
role in Total Safety Culture, 232
self-reinforcement need, 121
self-efficacy
    emotional intelligence and, 180
    empowerment and, 175–176
    strategies for enhancing, 184–185
self-esteem
    actively caring behavior and, 175
    coaching and, 144–145
    emotional intelligence and, 180
    strategies for enhancing, 183–184
    supportive conversation and, 144
self-fulfilling prophecy, 176–177, 184
self-help strategies for safety, 8–9
self-mastery, 62
self-motivation, *see* self-directed behavior
self-serving bias, 68, 94, 240
self-talk, 147, *see also* self-directed behavior
self-transcendence, 165–166
Seligman, M.E.P., 63
Selye, Hans, 59
Senge, Peter, 3
sensation, *see also* perception
    described, 44
    psychological dimension of, 38
signs, *see* poster/sign campaigns and safety
skill discrepancy in employees, 97
Skinner, B.F., 10, 17, 71, 76, 119, 161
Slovic, Paul, 49
small wins concept, 113, 184, 185
SMARTS goals, 117–118
social conformity and safety
    peer influence, 40–41, 226
    social support factors, 66–67
social loafing, 234
social responsibility norm, 167
Socrates, 93
SOFTEN, 142
SOON (Specific, Observable, Objective, and Naturalistic), 86
"start small and build" strategy, 191–192
states, person, *see* increasing actively caring, person state enhancement strategies
steering team, safety, 218–218, *see also* safety steering team
stimulus-response relationship, 76
stressors, *see* stress *vs.* distress
stress-related headaches, 57
stress *vs.* distress
    attributional bias, 67–68
    constructive and destructive aspects, 58–59
    key points, 60–61
    overview, 56
    person factors contributing to, 59
    stress definition, 57
    stress management, 5
    stressors coping mechanism
        examples, 62–63
        person factors, 62–63
        physical fitness, 63–64
        role in Total Safety Culture, 240
        social factors, 66–67
    stressors identification, 60–61

subjective well-being (SWB), 240
success seeker, 251
supportive conversations, *see also* communication in coaching; safety coaching
    conversation improvement considerations bias, 151–152
    flow direction, 148
    higher-level praise association, 153–154
    nondirective approach, 149
    personal approach, 153
    positive reinforcement and learning, 152–153
    power of conversation, 147
    praise in private philosophy, 154
    question-asking technique, 150
    receiving recognition, 155–158
    recognition of safety achievement emphasis on mistakes avoidance, 152–155
    safety and, 147
    second-hand recognition, 154–155
    self-image support, 145
    stand-alone recognition use, 154
    tangibles use, 154
    timing importance, 153
    transition from nondirective to directive, 150–151
    verbal commitment, 148–149
supportive feedback, 15, 101, 116, 120, 127, 136, 144–146, 152, 222, 229
supportive intervention, 100
surveys, 212, 213
synergy, 187–188
systems approach to safety, 28

T

talent, matching to jobs, 97
tangibles use in incentive/reward programs, 154
target audience involvement in interventions
    Airline Lifesaver, 113–114
    Flash for Life, 112–113
    ownership-involvement principle, 111
target behaviors definition, 84–85
    safe behavior promise example, 132
TEAM (Together Everyone Achieves More), 197
teams, safety
    ground rules for meetings, 199
    groups and belonging and, 187–188
    member selection, 197–198
    mission statement, 199
    paradigm shifts for, 28
    phases of teamwork
        action plan development, 200
        assignment clarification, 196–197
        charter establishment, 197–199
        consensus building, 197–199
        disband, restructure, or renew, 203–204
        evaluate performance, 202–203
        meeting procedures set, 200–202
TEDX talk, 244
Thank-You Cards, 130–131
Tharp, R.G., 103
theory-based safety
    behavior *vs.* person-based approaches
        behavior-based, 16
        cost effectiveness, 18–20
        described, 15–16, 20

elements of approach, 12
integration of, 19–20
mission statement development, 12, 15–16
person-based, 16–17
relevance to occupational safety, 14–15
theory as a map, 12–13
thinking skills and leadership, 223
threat, 60
timing and placement considerations in interventions, 114
Together Everyone Achieves More (TEAM), 197
total recordable injury rate (TRIR), 68, 186, 200, 207, 244
Total Safety Culture (TSC), *see also* culture change
    actively caring role in, 182
    aim of, 31
    critical behaviors identification, 82
    factors involved, 17–18
    leadership's role in, 43
    mission statement development, 12
    new three Es, 23–24
    paradigm shifts for
        accident proneness mentality, 6, 22
        achievement orientation need, 25–26
        behavior focused approach, 26–27
        bottom-up involvement, 27
        continuous improvement, 30
        corporate responsibility and, 24–25
        fact finding focus, 28–29
        interdependence, 27–28
        old three Es, 22–23
        paradigm definition, 24
        proactive, 30
        systems approach, 28
        value based, 30–31
    perception's relevance to, 48
    principles summary
        "all injuries are preventable" slogan, 238
        actively caring, 240, 241, 242
        bias and perception, 238
        bystander apathy, 242
        commitment and consistency, 243–244
        culture change requirements, 232
        DO IT process, 235
        empowerment, 243
        evaluation process, 238–239, 244–245
        example setting, 236
        external consequences in moderation, 236
        feedback importance, 235
        focus on receptive people, 232
        group-based behavior, 244
        human nature and safety, 234
        incentive program focus, 233
        instructor impact, 232
        internal change importance, 238–239
        internally driven safety, 231–232
        intervention impact, 235
        leadership and, 232
        learning new behaviors, 234
        motivation basis, 236
        negative consequence effects, 236–237
        perceived risk, 238
        personal control, 238
        power of authority, 234
        power of choice, 232
        reciprocity principle, 243
        risk compensation, 238
        safety coaching, 241
        Safety Triad, 233
        social loafing prevention, 234
        stressors coping mechanism, 240
        value based, 234
    tolerance for risk and, 53
    value of working safely, 31, 35
Tracey, Brian, 9
trained seal act perception example, 45
training, *see also* education
    assessing type needed, 98
    behavior-based, 99–100
    definition, 84–85
    overcoming resistance to change, 223–224
    vs. states, 174
traits improvement evaluation and, 212
Type A behavior *vs.* Type A emotion, 58
Type A personality, 57–58, 251
Type B personality, 58, 251

U

unconditioned response (UCR), 75, 76
unconditioned stimulus (UCS), 75, 79
unions and active caring, 188
upside-down consequences, 96
U.S. Department of Transportation, xxiv

V

values
    person factors and, 72–73
    role in Total Safety Culture, 234
verbal commitment, 149
vicarious consequences, 78
vicarious punishment, 78, 79, 251
vicarious reinforcement, 79, 111, 251
vision, 146, 156–157, 161–162, 180, 182, 230, 234, 251

W

warning beepers and habituation, 110
Watson, D.L., 103
Wilde, G.J.S., 53, 73
work underload, 60

Y

Yerkes-Dodson Law, 58, 59
Yerkes, R.M., 58–59

Z

Ziglar, Z., 9, 10, 117

9781032572611